Reviews of the Bestselling & Internationally Acclaimed CONSCIOUS HEALING: BOOK ONE ON THE REGENETICS METHOD ...

"In this paradigm-reworking book, author Sol Luckman develops the concept of 'Ener-genetics,' a synthesis of age-old and new-age wisdom with a sound- and light-based technology that has huge potential for human evolution and self-enlightenment ... This is revolutionary healing science that's expanding the boundaries of being."
—NEXUS NEW TIMES

"CONSCIOUS HEALING is one of the most important books I have ever read ... From Mr. Luckman's own personal, debilitating illness to discovery and transformation, we are brought to extraordinary levels of understanding."
—Andrea Garvey, Publisher, CREATIONS MAGAZINE

CONSCIOUS HEALING might "be the key that unlocks a new way of being."
—ODYSSEY MAGAZINE, Editor's Choice Book

"A welcome addition to personal, professional, and academic library Alternative Medicine reference collections, CONSCIOUS HEALING is to be given high praise for its remarkable coverage of the many intricacies of Regenetics and its progressive discovery, and is very highly recommended to all seeking an education on alternative self-healing procedures."
—Midwest Book Review

ALSO BY SOL LUCKMAN

NONFICTION

Potentiate Your DNA: A Practical Guide to Healing &
Transformation with the Regenetics Method

FICTION

Snooze: A Story of Awakening

HUMOR

The Angel's Dictionary: A Spirited Glossary for the
Little Devil in You

Conscious Healing

Book One on the Regenetics Method

Sol Luckman

Second Edition printed in 2010 by Crow Rising Transformational Media through Lulu Enterprises, Inc., 3101 Hillsborough Street, Raleigh, North Carolina USA 27607 **http://stores.lulu.com/solluckman** ISBN: 978-0-9825983-9-9

Library of Congress Control Number: 2009912716 Library of Congress, Cataloging in Publication Division 101 Independence Avenue, SE, Washington, DC 20540-4320

For those with ears to hear
& our not-so-little heron from Tula

CONTENTS

LIST OF ILLUSTRATIONS

PREFACE TO THE SECOND EDITION

When *Conscious Healing: Book One on the Regenetics Method* first appeared in 2005, my expectations (to the extent I had any) for this unprecedented blending of new science and "new age" spirituality were modest at best. I knew of no book quite like *Conscious Healing* then, and the same still applies today, so I literally had nothing to go on. And of course, self-publishing, while it can be rewarding, is a step into the unknown, with no support net to catch you if you fall.

So I was greatly surprised when *Conscious Healing* began putting up very respectable sales numbers indeed, consistently appearing on various bestseller lists at online venues such as Amazon.com, which it continues to do to this day.

I also was tremendously gratified by the enthusiastic editorial reviews *Conscious Healing* received, to say nothing of reader reviews that often spoke of a book that, by itself, was an "activation of consciousness," profoundly reminding readers of truths about human potential and conscious evolution they already knew but had forgotten, in whole or in part.

Nothing was so inspiring and humbling, however, as seeing so many readers go on to experience personal transformation through the Regenetics Method. While I affirmed from the beginning that *Conscious Healing* is a valuable book for its informational content alone, it hardly can be overemphasized that the life-changing effects of

Regenetics cannot be described fully in words or grasped on a purely intellectual level.

Having been involved in an editorial capacity with the Spanish translation of *Conscious Healing*, during which I noted some areas of the text I felt could use clarification and/or expansion, I already was considering creating a second English edition—when I was contacted by a European company wanting to release *Conscious Healing* in German. When our negotiations ended in a publishing contract, I took this as a sign, and immediately set about to update the entire book.

That is the text you have now. To those already familiar with the first edition of *Conscious Healing*, I propose that if you liked that version, you will love this one. I wish both my loyal and first-time readers to know that a lot of renewed energy has gone into describing and substantiating the Regenetics Method, as well as the evolutionary context that has fostered it, as clearly and accurately as possible.

I have added nearly twenty percent more text to the second edition, much of which incorporates leading-edge scientific and philosophical content. Realizing that readers might benefit from a more detailed explanation of the sequence of Regenetics activations, I also have included a description of the fourth and final phase of the Method, Transcension Bioenergy Crystallization, as well as a brand-new chart showing, at a glance, the progression of the Regenetics Method Timeline.

Readers interested in 2012, the Mayan calendar, ascension and related topics have not been overlooked. In particular, Chapter Nine, "The Shift in Human Consciousness," has been positively packed with new evidence supporting this book's thesis that humanity—individually as well as collectively—is poised on the brink of a thoroughgoing metamorphosis into a far more enlightened way of being.

Needless to say, in my process of revision I have expanded the book's already considerable references, which show up both in the main text and the Bibliography. This has had a measurable impact on the Index, which has grown proportionally. I also saw the need to add several new terms to the Glossary, and to refine the definitions of some preexisting terms to reflect my own evolving understanding of this uniquely empowering form of energy healing.

Finally, all three of the book's Appendices have been updated. Because they are both informative and inspiring, in Appendix A I share nearly twice as many Testimonials from clients. In Appendix B, I have added a number of Frequently Asked Questions and clarifications. And as for Appendix C, it is my sincere intention that readers will find the sample Electromagnetic Schematic more comprehensible on all levels.

In November of 2007, I wrote as editor in the introduction to a special issue of my popular free ezine, *DNA Monthly*,

> Repeatedly, I've felt my finger to be on the pulse of a global zeitgeist of transformation operating in and through the genetic level of consciousness [...] A consistent theme of feedback has been that a more expanded view of science and healing is not only welcome at this time; it is critically needed as we evolve out of a monolithic version of reality into the higher-paradigm understanding that there is no such thing as science—there are ultimately only *sciences*, any of which can have their strengths and weaknesses based on the strengths and weaknesses of those of who, consciously or not, create them. Now more than ever, it is incumbent upon us as a

species to create our science(s) of reality *consciously.*

In making this expanded and updated second edition of *Conscious Healing* available even as my novels on the power of the imagination gain greater recognition, I realize the extent to which my life's work has been to articulate both a science and an art of reality as consciously as possible. May *Conscious Healing* be a catalyst for you, as it has been for so many, in your inner and outer transformation.

—Sol Luckman
October 2009

INTRODUCTION

This is a truly exciting time to be alive. As a species, judging by our "postmodern" art and "subquantum" science, we are learning just how completely we create our own reality. Central to this evolution of human consciousness is a growing appreciation of the many ways we (collectively and individually) create ourselves. Literally. DNA is the alphabet we divinely endowed biological beings use to compose our existence.

Revolutionary new research in "wave-genetics" reveals DNA can be activated—noninvasively—by radio and light waves keyed to human language frequencies. Studies by cell biologists further demonstrate that the genetic code can be stimulated through human consciousness—specifically, the unity consciousness associated with unconditional love—to heal not only the mind and spirit, but the body as well.

Benefits of DNA activation can range from allergy relief and heightened energy, to healthier relationships and increased abundance, to personal transformation and renewed life purpose. Since DNA regulates all physical, mental, emotional and spiritual aspects of our being, the possibilities are endless!

In the words of bestselling author and leading health researcher Dr. Leonard Horowitz, DNA with its "stunning ability ... to function as a receiver and transmitter of ... the Divine Cosmic Song" represents "far more than a static blueprint for body building." DNA, which Horowitz calls the "Sacred Spiral," is the "ideal super conductor, and micro-antennae, designed to perfection beyond the reach of the

wildest imagination, for physical re-spiritualization."
DNA constitutes the "central processing station for
human electricity and evolutionary destiny. If
'knowledge is power,' revealing [DNA's] secrets is the
key to personal empowerment, spiritual evolution,
and even planetary salvation."

Taken together, such pioneering, relatively
unknown, and often suppressed genetic research
reveals an astonishing wealth of potential: to "reset"
bioenergetic systems damaged by trauma and
toxicity, to stimulate life force and creativity, even to
"switch on" untapped capabilities in the human brain
as steps toward unity consciousness and its
corresponding evolved biology. Thanks to a holistic
technique for DNA activation we at the Phoenix
Center call the Regenetics Method, an affordable,
effective means is now available to "potentiate" one's
entire being.

Potentiation Electromagnetic Repatterning, a
central focus of this book, is the first of four
integrated DNA activations that make up the
Regenetics Method. Regenetics is a unique synthesis
of ideas in action designed to improve lives in
numerous ways, tangible and subtle. It is my intent
that your own life will be improved even as you
thoughtfully consider the concepts presented herein.

Additional information on the second, third
and fourth phases of the Regenetics Method—
Articulation Bioenergy Enhancement, Elucidation
Triune Activation, and Transcension Bioenergy
Crystallization, respectively—is available online
at **http://www.phoenixregenetics.org** and
www.potentiation.net.

This text is divided into two principal parts,
the first reflecting the "micro" aspect of Regenetics,
the second providing a "macro" perspective on this
Method in relation to human evolution. In the first
part, "Textual Healing with Potentiation

Electromagnetic Repatterning: Introducing the Art & Science of the Regenetics Method," I share the story of my nearly eight-year chronic illness, with an emphasis on the development of Potentiation and how I used this DNA activation to heal myself.

Part II, "Sacred Cosmology, Sacred Biology: The Regenetics Method & the Evolution of Consciousness," complements Part I with a macrocosmic presentation of Regenetics in the greater context of the evolution of consciousness. Here I draw on the work of many outstanding thinkers from a variety of orientations who share the belief in a purpose and direction to human life—that as a species we are evolving in a specific manner on a prescribed timeline. It is my intention that Part II will stimulate your curiosity about human evolution in general while inspiring you to explore avenues such as the Regenetics Method for fully actualizing your own unique potential.

In addition to Parts I and II, I have included three substantial Appendices as educational resources. Appendix A provides wide-ranging Testimonials from individuals who have experienced one or more phases of the Regenetics Method. Appendix B contains detailed answers to Frequently Asked Questions about Potentiation and the Regenetics Method. Appendix C is a representative Electromagnetic Schematic of the first of twelve "Electromagnetic Groups" encountered during the development of the Regenetics Method as described in Chapter Four.

Additionally, I have provided an extensive Index as well as a comprehensive Glossary of Terms derived from various disciplines that may be new to the reader or that I employ in specialized ways. I invite you to use this Glossary as a reference and also to read it in its entirety as a "journey in consciousness." Finally, a complete Bibliography is included to assist those who wish to explore the science and philosophy behind Regenetics in greater depth.

Many of the concepts behind Potentiation Electromagnetic Repatterning and the Regenetics Method outlined in this book may strike the reader as "cutting-edge." While they indeed may appear that way in this day and age, most of them also are, paradoxically, extremely old.

These concepts stem from time-honored practices (that are truly *practical*) rooted in the curative power of prayer, shamanic medicine, and specifically the balanced use of *sound* and *intention* to heal in ways that can seem miraculous to many Westerners. I will share some fascinating scientific research that substantiates these phenomena. But before proceeding, allow me to call your attention to the old adage of how a new truth typically comes to be accepted.

"All truth passes through three stages," wrote Arthur Schopenhauer. "First, it is ridiculed. Second, it is violently opposed. Third, it is accepted as being self-evident." In a similar vein, Albert Einstein once said, "If at first an idea isn't absurd, there's no hope for it."

I offer that the Regenetics Method (and modalities based on similar principles) represents one such "absurdly profound" truth that eventually will become self-evident. Thank you for your interest in this work and willingness to expand—perhaps—your vision of who you are and what is humanly possible.

—Sol Luckman
Cofounder, Phoenix Center for Regenetics
Facilitating conscious personal mastery as a bio-spiritual healing path through integrated DNA activation.

Editor, *DNA Monthly*
Your FREE online resource for cutting-edge news about who you truly are.
http://www.phoenixregenetics.org
http://www.potentiation.net
December 2005

NOTE ON TERMINOLOGY

The following text, like the Regenetics Method, is a far-reaching synthesis of concepts from many disciplines running the gamut from hard science to speculative cosmology. Such a comprehensive approach, while demonstrating the considerable scope of Regenetics, is not without presenting certain difficulties for some lay readers— particularly at the level of terminology.

Time and again, my multidisciplinary research has revealed that the numerous fields of scientific and philosophical inquiry have produced different terms for describing the same phenomena. While this tends to validate the phenomena themselves as essential to human experience, confusion can result unless we understand that many of the words and phrases employed by the various disciplines are, for practical purposes, synonyms. Wherever appropriate, I have provided definitions in the Glossary for such cross-referencing terms that highlight both their similarities and nuances.

Two key examples should suffice to call attention to this dynamic and facilitate understanding. The following terms for the apparently dormant and unused portion of DNA are, for the purposes of our discussion, virtually synonymous: "junk" DNA, "jumping DNA," potential DNA, introns, and transposons. This subject is covered in detail in Part I.

Similarly, as elaborated in Part II, torsion radiation or energy is simply a scientific way of conceptualizing the conscious, generative force of unconditional love. Both originate from Galactic Center (also referred to as Source, Tula, the Logos, and the Healing or Central Sun) as the "tone" of Ge, which differentiates into spiral standing waves of sound and intention (light).

Other words and phrases used to describe aspects of torsion energy or unconditional love include: Silent Stillness, aether, universal creative consciousness, thought, tachyons, scalar waves, life-wave, orgone energy, chi, prana, and kundalini.

"In the order of healing, it is human consciousness that first must change."

—Ken Carey, *Return of the Bird Tribes*

Conscious Healing

• PART I •
TEXTUAL HEALING WITH POTENTIATION ELECTROMAGNETIC REPATTERNING:
INTRODUCING THE ART & SCIENCE OF THE REGENETICS METHOD

1
Nonlocalized Mind & Era III Medicine

One of many inspirational figures behind Potentiation Electromagnetic Repatterning and the Regenetics Method is Larry Dossey. The reader may be familiar with his bestselling book *Healing Words*, an account of the therapeutic effects of prayer substantiated by numerous scientific studies performed at highly respected academic institutions such as Harvard and Stanford. In *Reinventing Medicine: Beyond Mind-body to a New Era of Healing*, Dr. Dossey reiterates:

> Many studies reveal that healing can be achieved at a distance by directing loving and compassionate thoughts, intentions, and prayers to others, who may even be unaware these efforts are being extended to them. These findings reveal the ability of some part of our mind or consciousness to escape its confinement to the brain and body and to act anywhere, regardless of distance.

An important point emerges from this groundbreaking consciousness research: there appears to be no single "correct" type of prayer. "These studies clearly show that *healing intention* is a general term," Dossey emphasizes. "It can be secular or religious; it may or may not involve prayer."

Christians praying to their God for healing, for example, were no more or less successful than Muslims or Jews praying to theirs. So statistically eye-opening were these studies many hospitals now offer nondenominational prayer for patients undergoing life-threatening surgeries. Survival rate and recovery speed are enhanced beyond any doubt by these noninvasive intercessions that have the added merit of being cost-effective.

Since the publication of *Healing Words,* a variety of additional studies on prayer and healing have been performed, including an impressive double-blind analysis performed at UCSF-California Pacific Medical Center that showed positive effects of prayer on patients with advanced AIDS.

Other research on the healing power of prayer suggests that a major determining factor of success or failure is the level of *nonattachment* of the pray-er. Between 1975 and 1993, the Spindrift Foundation performed hundreds of thousands of tests to assess the effectiveness of directed prayer (i.e., focused on a specific outcome) versus non-directed prayer (in which only what is best for the person is requested). Both directed and non-directed prayer worked better for the control group for whom no prayers were known to be said, but non-directed prayer showed a significantly higher success rate than directed prayer.

In *The Isaiah Effect: Decoding the Lost Science of Prayer and Prophecy,* bestselling author Gregg Braden takes the practical applications of prayer a step further. Basing his claims on his reading of one of the Dead Sea Scrolls known as the Isaiah Scroll, Braden describes how the Essenes from the time of Christ employed a type of prayer designed to affect, and effect, quantum outcomes by literally changing the pray-er's picture of reality. This form of "active prayer," to which I will return in Part II, is a method of focusing intention that validates whatever one is praying for as already having been granted.

"Rather than creating or imposing change upon our world," theorizes Braden, "perhaps it is our ability to change our focus that was the ancient key suggested by the masters of passive change in history" such as Buddha, Gandhi, and Jesus. "Quantum physics suggests that by redirecting our focus—where we place our attention—*we bring a new course of events into focus* while at the same time releasing an existing course of events that may no longer serve us." Author John English, whose novel *The Shift: An Awakening* won the Coalition of Visionary Resources Book of the Year Award in 2004, speaks in similarly compelling terms about the powerful human ability to "dream a new world into being."

> *Active prayer is a method of focusing intention that validates whatever one is praying for as already having been granted.*

In *Reinventing Medicine*, Dossey, the former chief of staff at a major Dallas hospital, examines allopathic approaches to healing in light of the principle of "nonlocality" often discussed in relation to quantum physics. Putting modern medicine in quantum perspective, Dossey admits that we "are facing a 'constitutional crisis' in medicine—a crisis over our *own* constitution, the nature of our mind and its relationship to our physical body."

To help explain this "constitutional crisis," and to assist humanity in transitioning beyond it, Dossey outlines three eras in the history of Western medicine. While these eras are by no means mutually exclusive, each has a dominant, defining focus.

The first era began in the 17th century with Cartesian thinking and was characterized by a mechanistic view of the body. Era I medicine looks at the human body more or less as a machine that can be manipulated. In this rather primitive medical model, there is no place for mind or consciousness—

and certainly none for "soul" or "spirit." Surgery and vaccines are both applications of Era I medicine.

In the 19ᵗʰ century, according to Dossey, Era I gave way to Era II with the acknowledgement of the so-called placebo effect. Era II saw the birth of psychoanalysis and psychiatry and is characterized by mind-body approaches to healing.

Era II medicine is based on the observation that your mind and body are connected intimately such that *your* consciousness can remedy *your* physiology in provable ways. This is the "power of positive thinking," as Norman Vincent Peale phrased it. Era II techniques continue to play an important role in today's medical paradigm.

As a global culture, we now are in the process of greatly expanding that paradigm with what Dossey refers to as Era III medicine, also called "nonlocal." The hallmark of Era III medicine is the "nonlocalized mind," meaning the universal Mind that also has been called unity, Christ, Buddhic and even God consciousness. Physicists sometimes conceptualize nonlocalized mind as the "Unified Field," psychologists often refer to it (following Carl Jung) as the "Ground of Being," while many spiritualists speak of "Source," to borrow a term popularized by famed psychic and healer Edgar Cayce in his remarkable readings.

The term *nonlocal* is particularly relevant to this discussion because of its derivation from science. Physicist David Bohm, one of the founders of the holographic model of reality, has used the phrase "quantum potential" to refer to the nonlocal point in space where space ceases to exist and two electrons, for instance, can occupy the same coordinates. Nonlocalized mind has been described in various ways, but simply put, it is the Supreme Consciousness of which we are all facets—regardless of whether we choose to acknowledge our interconnectedness.

The fundamental notion behind Era III medicine, very evident in the writings of Braden and English, is that the human mind can operate *outside* the confines of the physical body and positively impact other people, animals, and even the environment.

Nonlocalized mind has been described in various ways, but simply put, it is the Supreme Consciousness of which we are all facets— regardless of whether we choose to acknowledge our interconnectedness.

Some may be conditioned to believe this is impossible. Others may be aware of such studies as Princeton Engineering Anomalies Research (PEAR) that present solid, empirical evidence of the ability of human consciousness to change physical reality, including the mind's capacity to affect the outcome of random-number generators and alter the rate of radiation emissions as measured by a Geiger counter.

I offer this first-person narrative of the development of Potentiation Electromagnetic Repatterning to members of both these groups and all others willing to open their minds and hearts to the essential truths that underwrite the Regenetics Method—truths that can be applied productively in many areas (personal and professional) besides DNA activation.

In the following pages, I will focus my observations primarily on the positive impact of nonlocalized mind on human beings, but the majority of the concepts I introduce easily should be understood as applying to the world at large.

2
Autoimmunity & Energy Clearing's Brave New World

Before further exploring the pivotal concept of Era III medicine, nonlocalized mind, a word or two about my own background is in order. I grew up in a small Southern town, where I excelled athletically and academically. I was quarterback as well as valedictorian and ended up winning a prestigious college scholarship.

At almost exactly the same time, I was offered the chance to play Division I football—which I declined because I already had academic funding. I went on to graduate at the top of my scholarship class, win a Fulbright teaching award, and receive an Ivy League doctoral-student fellowship followed by two national research grants in the humanities.

My most remarkable characteristic as a younger man was an intense passion to experience life fully. The implications of this passion only dawned on me after living "fully" came to include suffering disease and descending into a "dark night of the soul" after the onset of a mysterious, debilitating illness in 1996. At that time, I was working on a Ph.D. in literature. This is significant because as a novelist, I tend to approach the Regenetics Method from a "literary" perspective, as the titles of this section and chapter suggest. I will elaborate momentarily.

To make a very long story short, at twenty-seven life inexplicably came crashing down. One day I was exercising three hours at a stretch, able to eat and drink whatever I pleased. The next I was gripped by a mysterious illness that, one by one, took away the foods, drinks and sports I loved, even—in the insidious way chronic illness has of stripping one clean—many people I loved.

Despite a string of "negative" medical tests, I lay in bed night after night terrified I was dying. In addition to debilitating allergies and exhaustion, I suffered from approximately thirty seemingly unrelated symptoms, including back pain, hypoglycemia, receding gums, skin rashes, shortness of breath, muscle twitching, and chemical sensitivities.

In an effort to halt my deterioration, I took nearly every supplement on the market. I received regular intravenous chelation. I experimented with ozone and infrared saunas. I went on parasite cleanses, special diets for Candida. I tried reiki, acupuncture, homeopathy, biofeedback, magnets, "zappers." I underwent EMDR, hypnosis, radionics, even "psychic surgery." But after trying practically everything and spending a fortune, I was sicker than ever and getting worse.

Eventually, my health deteriorated to where I could only eat meat and vegetables. At one point, my white blood cell count was alarmingly low. I never was diagnosed officially with anything but the medical profession's catchall for baffling conditions— "depression"—although I now am certain my disease was of a serious autoimmune nature precipitated by hepatitis and yellow fever vaccines I received before traveling to South America for dissertation research in 1995.

Even at the time of my diagnosis, I felt deeply that depression was the result not the cause of whatever was degenerating my once-athletic body. But I dutifully popped my pills until I nearly died of an adverse reaction. Growing desperate, with little left to lose, I headed into "alternative" territory seeking solutions.

***

In terms of understanding what had gone haywire in my physical body, my biggest breakthrough came when I grasped the role vaccines play in creating autoimmunity. For a sobering look at the potentially disastrous consequences of vaccines, I recommend Leonard Horowitz's *Emerging Viruses: AIDS and Ebola—Nature, Accident or Intentional?*

In this eye-opening bestseller, Harvard-trained Dr. Horowitz persuasively argues that vaccines are the root cause of a long list of autoimmune diseases, including AIDS. Following the discovery of simian 40 retrovirus (sometimes referred to as "monkey AIDS") in polio vaccines, many other researchers, among them dozens of biologists and medical doctors, have reached similar conclusions.

After a year spent testing Horowitz's ideas, I concluded that immune-wrecking retroviruses can penetrate the bloodstream via "immunizations" and alter one's genetic code, potentially sabotaging health under a myriad of creative diagnoses such as "fibromyalgia," "chronic fatigue," and "multiple chemical sensitivity."

This may strike anyone who accepts the official line that vaccines are safe and effective as unbelievable. But after a year spent testing Horowitz's ideas at the energetic level, I concluded that immune-wrecking retroviruses can penetrate the bloodstream via "immunizations" and alter one's genetic code, potentially sabotaging health under a myriad of creative diagnoses such as "fibromyalgia," "chronic fatigue," and "multiple chemical sensitivity." Many thousands of unsuspecting people have died or been handicapped permanently following adverse reactions to vaccines. Even the medical establishment recently linked certain childhood vaccines to autism.

A major contributing factor to many, if not all, autoimmune conditions is genetic damage through invasive factors such as vaccines compounded by cellular toxicity. I contend that cells collect and hold toxicity for the purpose of slowing down the many mutant pathogens, such as simian 40 retrovirus, released in the organism under the radar of the immune system by vaccines.

The body knows that toxic substances—heavy metals and pesticides, for instance—not only are poisonous to the host, but also to pathogens. Such a Catch-22 can lead to environmental illness and immunological breakdown in which the body starts attacking its own toxic cells, but it may be the only choice a biosystem operating with damaged DNA has.

Many people are led to believe that since they have an autoimmune disorder, they are more toxic *because* of their chemical, environmental or nutritional sensitivities. Another way of saying this is that it commonly is assumed the body becomes more toxic in autoimmune states because it cannot or does not know how to detoxify.

Based on my research and personal experience of genetic collapse, however, it appears that autoimmunity in large measure is induced by foreign genetic invaders (which can include genetically modified foods, or GMOs) that negatively reprogram DNA by utilizing the body's RNA transcription process, instructing the body to replicate artificial codes inside cells. In other words, once DNA is reprogrammed, it literally has the ability to grow new pathogenic—perhaps "pathogenetic" would be a better word—cellular cultures.

According to Horowitz and many other researchers, vaccine-induced pathogens, in addition to simian 40 retrovirus, can include prions, mycoplasmas, mouse parotid tumor tissue, bovine lymphotropic virus, feline leukemia virus, Epstein-Barr virus, and Rous sarcoma virus—to name only a few. When these are "uploaded" into the genetic code using the reverse transcriptase ("backward writing") enzyme, any number of autoimmune conditions can result—from lupus to leukemia, depending on the individual's constitution and lifestyle and the number and type of vaccines received.

> *The body, in its wisdom, realizes it has been altered fundamentally, but like a computer it must carry out the codes in its reprogrammed DNA. This can lead to a degenerative defense response as the body accumulates more and more toxicity in an attempt to "short-circuit" the foreign pathogens being grown like weeds in the cells. The body simply uses what is available from the environment in its biological war against itself.*

The body, in its wisdom, realizes it has been altered fundamentally, but like a computer it must carry out the codes in its reprogrammed DNA. Over time, this can lead to a degenerative defense response as the body accumulates more and more toxicity in an attempt to "short-circuit" the foreign pathogens being grown like weeds in the cells. The body simply uses what is available from the environment in its biological war against itself.

What often happens with Candida following such genetic damage is very telling. Contrary to popular belief, there is nothing inherently wrong with Candida. In a properly functioning body, *Candida albicans* keeps tissues healthy by scavenging potentially harmful microorganisms and toxins.

Candida only gets out of control when the body tries to defend itself from some other invasion, usually of a genetic nature. This cycle is nearly impossible to halt without interceding "energenetically" because the problem is in the DNA, which unless directed to resume normal biological operations, continues a vicious cycle of replicating its mutated codes, then futilely trying to clean up microorganism overgrowth with more overgrowth!

Sensitivities and allergies result when the body is so occupied with the war going on at the level of the cells and immune system it cannot handle or take on additional foreign substances. In many cases, microorganism populations are so out of balance they actually consume the host's food and produce additional toxic waste, such as fungal mycotoxins, inside and outside cells—further exacerbating an already genetically entrenched state of autoimmunity.

In the second year of my illness, in consultation with a holistic doctor I became convinced mercury poisoning was a factor deranging my immune system. After reading an enormous amount of material on the controversial subject of mercury and other heavy metals in dentistry, I decided to undergo the painful process of having my amalgam fillings removed following a protocol similar to one developed by Dr. Hal Huggins. Actually, I ended up going through this protocol *twice*, after my first dentist made a mistake and replaced my mercury with barium and other metals![1]

---

[1] As it turned out, much of the material I read on mercury cross-referenced the vaccine issue. "We are changing our genetic code through vaccination," write health researchers Robert and Kerrie Broe, to cite one example. In the future, the Broes insist, basing their position on

For all the trauma it caused my mouth (and wallet), getting my dental work redone and removing such a toxic load of heavy metals from my system sparked a temporary "spontaneous remission." I was immediately almost back to my old self.

I went from being unable to eat carbohydrates or drink alcohol, before removal of my amalgams, to getting up out of the dental chair after my fourth and final appointment and being able to enjoy pizza and beer that afternoon! I had no Candida or leaky gut reaction whatsoever. To one who had been allergic to nearly everything, it was a miracle. I was able to resume my normal routine for about five months—until I began to lose ground again.

This time I rapidly went downhill. In desperation, I left graduate school and moved to New Mexico to study qigong. Qigong is an ancient Chinese technique of energy healing related to tai chi. I studied with a master who had cured himself of chronic fatigue syndrome (CFS or CFIDS) while in his twenties. I practiced at least three hours a day for a year while receiving weekly acupuncture treatments and taking copious quantities of Chinese herbs.

This probably saved my life and certainly got me back on my feet. I took a teaching job at a preparatory school and actually was strong enough to go trekking around in the woods with a heavy backpack. But I was still having regular food reactions. And if I went more than a couple days without my intensive regimen of qigong, I started to re-experience old symptoms that included involuntary muscle spasms, tingling in my extremities, facial neuralgia, and other frightening

---

decades of in-depth research, it will be acknowledged that two of the "biggest crimes against humanity were *vaccines* and *mercury dental fillings*."

neurological sensations. At various points during my illness, I was convinced I had multiple sclerosis (MS) or Parkinson disease.

Then of all foolhardy things, I accepted a teaching position at an international school. During my first and only year, I developed back-to-back abscessed teeth that required antibiotics.

This was the last straw. The antibiotics wrecked my fragile biological terrain (see Glossary) and I was down for the count. I slept fifteen hours a day, had to quit my job, and went from an already severely restricted diet to being unable to season my meat and vegetables because of crippling allergic reactions. I once quipped during that nightmarish period after returning to the United States with my tail between my legs that I could eat practically anything—as long as it was not food.

It was at this point, when hope was failing and I suspected I was dying, that I stumbled on the brave new world of energy clearing. I am referring specifically to NAET, Nambudripad's Allergy Elimination Technique, and an offshoot of this therapy called BioSET, developed by one of Dr. Nambudripad's students, a chiropractor named Ellen Cutler.

As soon as I started receiving NAET and BioSET treatments, I noticed some positive physical shifts. My food allergies temporarily subsided, although they never disappeared entirely, and I experienced some much-needed detoxification. The shifts were so immediate and palpable I found myself considering the possibility of using these modalities to heal my genetic damage, detoxify my cells, and rebuild my deteriorated tissues.

NAET derives in part from the homeopathic discovery that energy signatures can be imprinted in small glass vials using an electro-acupuncture device. For example, one can place the energy signature (a particular frequency the body will recognize) of an allergen such as sugar in a vial containing pure water and a drop of alcohol. The immune system's response to the vial is identical, for practical purposes, to its reaction to sugar. Although their cause never is explained adequately, allergies are seen in NAET as chemical, environmental or nutritional sensitivities that tend to derange the immune system, contributing to a variety of chronic ailments.

The patient then holds the vial with the allergen's energy signature while the practitioner performs acupressure along the spine designed to initiate a "clearing" using the nervous and Eastern meridian systems (see Glossary). The basic idea, similar to that of acupuncture, is to eliminate "blockages" that keep life force or bioenergy from flowing properly through the body.

In theory, clearings retrain the body, specifically the immune system, to accept substances formerly rejected as allergens. The popularity of NAET even among some members of the allopathic community attests to the fact it can produce measurable benefits.

BioSET expanded on NAET by recognizing that if it is possible to clear with one vial at a time, it should be possible to clear using multiple vials. One can clear sugar allergies with *Candida albicans*, which can feed on sugar, and even add vials that represent the pancreatic system, since insulin has a close relationship with sugar. Theoretically, it is even possible to clear heavy metals, viruses and other pathogens that might be impairing pancreatic function. BioSET has evolved to such a level of complexity that instead of vials, many practitioners now employ computers for clearings with specialized software and equipment designed to introduce multiple combinations of homeopathic frequencies to the patient.

Initially, I was tremendously encouraged by this approach—especially after I saw some of my severe allergic patterns improve. I ended up pestering my practitioner until she trained me in her own version of BioSET and I began making my living with this therapy. I did this for a little over a year, during which I continued to receive treatments from my teacher, before I grew frustrated at my lack of progress and started treating myself.

In order to "reset" our bioenergy blueprint in a way that not only will "take," but also "hold," we must go directly to the source of the electromagnetic malfunction in our genetic code. Only in this manner can we reestablish the energetic harmony and coherence necessary for sustained wellbeing. In order to do this, it is necessary to employ sound along with intention to activate the self-healing mechanism in the apparently unused portion of DNA.

In total, I received approximately seventy NAET and BioSET treatments. Doing the math, 70 × $75 at the going rate, I received over $5,000 in treatments. But if anything, after a brief plateau I was going downhill once again. I was becoming more and more fatigued, losing the foods I partially had gotten back, and even experiencing a variety of new symptoms.

Traditional energy clearings work by way of the nervous and meridian systems. But geneticists have begun to refer to DNA, not the nervous system, as our "biocomputer." In order to "reset" our bioenergy blueprint in a way that not only will "take," but also "hold," we must go directly to the source of the electromagnetic malfunction in our genetic code.

Only in this manner can we reestablish the energetic harmony and coherence necessary for sustained wellbeing. Ultimately, I concluded that in order to do this, it is necessary to employ *sound* along with *intention* to activate the self-healing mechanism in the apparently unused portion of DNA. But first, it was necessary to understand the critical role played by the body's auric or electromagnetic fields in creating disease and maintaining health.

3
The Electromagnetic Fields: Our Bioenergy Blueprint

My last several months working with an offshoot of BioSET, I began intuitively receiving a tremendous amount of information. At this stage, I was inspired to read a book nearly every other day on biology, chemistry, physics, genetics, new science, energy medicine, or esoterics.[2] A host of interrelated ideas came to me all at once and I began to substantiate them through muscle testing.

The science of muscle testing (kinesiology) employs muscle response (strong or weak) tests to determine allergies, emotional blockages, and even the truth or falsehood of given statements. Since its invention in the 1960s, kinesiology has become popular among both alternative and mainstream healthcare professionals around the world.

For those unfamiliar with the many applications of kinesiological testing, I recommend *Power vs. Force: The Hidden Determinants of Human Behavior* by Dr. David Hawkins. This famous text, praised by Mother Teresa among others, provides a thought-provoking introduction to the fascinating field of kinesiology. For present purposes, it is simply necessary to mention that muscle testing, properly utilized, can be a powerful tool for gathering and evaluating information that has been validated empirically on numerous occasions.

---

[2] A number of these books are listed in the Bibliography.

My lovely partner Leigh assisted me throughout the development of the Regenetics Method. Without her keen insight and unwavering support through the ups and downs, this work never would have come into being.

We performed hundreds of hours of muscle testing—literally tens of thousands of tests—with ourselves and our clients. We would come up with an idea and order the homeopathic vials to assess it from a vial maker on the West Coast. He must have thought we had a few screws loose because we ordered vials for auric fields, *chakras*, sounds, notes, octaves, letters of the alphabet. But we paid him and he sent us the vials. Then we would test and retest enough to determine whether our ideas had any validity.

One of our most important early realizations about traditional energy clearings such as those used in ® and BioSET was that these techniques employ a typically "Western" focus on the physical—even though the techniques themselves use pure energy! In light of this internal contradiction—which was so obvious that at first, like trees in the forest, it was difficult to see—we became interested in the body's energy fields: specifically, the auric or electromagnetic fields.[3]

---

[3] I am choosing to use the adjective "electromagnetic" to characterize the levels of the human bioenergy structure, knowing full well that *electromagnetic* is more descriptive than scientific. These levels are actually "torsion" fields of conscious energy associated with the temporal dimensions that give rise to such effects as electromagnetism. But since it is difficult for a first-time reader to connect with a new word like *torsion* (which I will revisit), and since "auric" may sound too esoteric for some, I will stick with electromagnetic most of the time.

Figure 1: The Human Bioenergy Blueprint
From the perspective of quantum biology, the human
body is a hologram composed of intersecting lines of
bioenergy. The above figure shows how the vertical,
light-processing *chakras* **interface with the horizontal,**
sound-generated electromagnetic fields to create the
geometric matrix necessary for physical manifestation.

The electromagnetic fields can be thought of as
an interlocking set of high-frequency "force-fields,"
each responsible for the correct functioning of a
particular gland, meridian, organ system, set of
emotions, etc. Throughout this text, I will focus
attention on the electromagnetic fields. It should be
remarked, however, that as the chakras (see
Glossary) align with these fields in order and
number, many of the same observations also may be

applied to the chakras. The electromagnetic fields, combined with the system of chakras, form the human bioenergy blueprint that can be envisioned as an energetic grid—a hologram—of intersecting horizontal and vertical lines of force (Figure 1).

Many researchers have confirmed the existence of the human electromagnetic fields. Kirlian photography has captured these fields for decades. In the 1980s, Dr. Hiroshi Motoyama, a Japanese scientist, developed instrumentation capable of measuring bioluminescent electromagnetism such as light emitted from the chakras of yoga masters.

Valerie Hunt, a professor at UCLA and author of *Infinite Mind: Science of the Human Vibrations of Consciousness*, has employed an encephalograph (EEG) machine successfully to register the bioenergy fields. Dr. Hunt goes so far as to theorize that the mind, rather than residing in the brain, actually exists in the electromagnetic fields—and that in some as yet poorly understood way, the latter may *be* the mind.

In the main, this is consistent with other scientific research indicating that the mind clearly transcends the physical boundaries of the brain. Recent studies in immunological function conducted by Robert Jahn and Brenda Dunne, for example, concluded that mind or consciousness operates at the level of the immune system, which is directly responsible for distinguishing self from other.

From a more esoteric perspective, gifted psychic and author Sheradon Bryce states point-blank: "Your mind is your electromagnetic field, your auric bands, or whatever you wish to call them. Your field is your mind. The thing in your head is your brain." Interestingly, anticipating our discussion of the "ener-genetic" similarities and distinctions between sound and light that begins in Chapter Six,

Bryce refers to the electromagnetic fields as "feeling *tones*" (my emphasis).

The Jewish alchemical science of the Kabala calls the auric fields collectively the *nefish*, often described as an iridescent bubble surrounding the body. In the book *Future Science*, John White and Stanley Krippner point out that nearly a hundred different cultures refer to the human aura with nearly a hundred different names. The aura even may appear as a halo around medieval images of Christian saints. One reason Western science has ignored the aura is that, because of its extremely high (hyperdimensional) frequency bands, it is hard to quantify. But it is worth noting that most scientists fail to understand the true nature of *any* energy field, not even one as mundane and measurable as electrical current.

As our bioenergy blueprint, the electromagnetic fields function as a compendium of all the data pertinent to our wellbeing. In *The Holographic Universe*, Michael Talbot explains, "Because an illness can appear in the energy field weeks and even months before it appears in the body, many ... believe that disease actually originates in the energy field. This suggests that the field is in some way more primary than the physical body."

Naturopath Stephen Linsteadt, author of *The Heart of Health: The Principles of Physical Health and Vitality*, explains that an "interruption or distortion in the range, strength and coherency of the body's electromagnetic system leads to breakdown in the body's self-healing mechanisms." Physician Richard Gerber, author of *Vibrational Medicine*, goes a step further by arguing that if doctors could find a way to treat the bioenergy field, they would achieve total healing. Until then, Dr. Gerber contends, many treatments "will not be permanent because we have not altered the basic [blueprint]."

Similarly, Nataliya Dobrova describes the individual as a "complex emotional bio-energy information system: a microcosm that reflects a macrocosm—the universe. All of a person's organs and systems have their own electromagnetic rhythms. Disharmony in this rhythmic activity signifies disease." Dr. Dobrova goes on to explain how such an "imbalance is closely connected with structural or functional problems found in a person's organs or systems. If one can restore the person's own rhythmic harmonies to a sick organ, one can restore the proper functions of that organ."[4]

All manifestations of disease, whether "physiological" or "psychological," result from disruption of the primary ener-genetic harmonies and rhythms contained in the electromagnetic fields and corresponding chakras. These bioenergy centers have an intimate relationship with DNA that gives them direct regulatory access to all cellular functions. If we can find a way to reset this bioenergy blueprint through harmonic resonance, we can go directly to the root of disease.

A nearly identical line of thinking informs one of the classics in the field of sound healing, Jonathan Goldman's *Healing Sounds: The Power of Harmonics*. Through harmonic resonance, writes Goldman, "it is possible to restore the natural vibratory frequencies of an object that may be out of tune or harmony. When an organ or another portion of the body is vibrating out of tune, we call this 'disease.'"

Such belief in the power of harmonics to heal the body is echoed by Horowitz, whose research in

---

[4] Quoted in Joe Champion, "Transdimensional Healing with the ADAM Technology."

cymatics (the study of the effects of sound on physical form) and electrogenetics leads him to emphasize that "harmonic frequencies maintain health, promote growth and healing, while discordant frequencies produce stress, oxygen deprivation, acidification, electrochemical imbalances, illness and death."

From a cymatic or vibratory standpoint, *disharmony is disease.* The critical concept to grasp here is that all manifestations of disease, whether diagnosed as "physiological" or "psychological," result from disruption (in the form of toxicity or trauma, or both) of the primary ener-genetic harmonies and rhythms contained in the electromagnetic fields and corresponding chakras.

As I will elaborate in subsequent chapters, these bioenergy centers have an intimate relationship with DNA that gives them direct regulatory access to all cellular functions. Therefore, if we can find a way to reset our bioenergy blueprint through harmonic resonance, we can go directly to the root of disease processes. This was one of the principal goals Leigh and I set for ourselves as we began to develop the Regenetics Method.

4
Mapping the Bioenergy Fields

Having spent nearly eight years dying, I am deeply grateful for the pioneering work of Drs. Nambudripad and Cutler, without whom I do not know where (or even if) I would be today. Their inspiring techniques served as an indispensable springboard for the development of the Regenetics Method. But here I must point out two major blind spots with traditional energy clearings, at least as vehicles for resetting the body's electromagnetic blueprint.

The first oversight, to reiterate, is a predominant focus on physical issues without fully acknowledging their origins in our bioenergy fields. The second problem with traditional energy clearings is that the nervous system simply cannot process all the frequencies encoded like radio waves in our electromagnetic structure so as to transform a damaged blueprint.

The same shortcoming applies to most—otherwise beneficial—energetic modalities, such as reiki and radionics, which typically function at the comparatively "surface" level of the nervous system as opposed to through DNA. Another way of stating this, one I trust will become clearer as we proceed, is that the majority of energetic therapies in existence today are "light-based," lacking the genetically transformational aspect of *sound* (Figure 9).

Contrary to the conservative paradigm that insists healing must be achieved "one baby step at a time," my own experience and observation suggest that chronic illness in particular requires a *radical, simultaneous* bioenergy reset—one that can be accomplished only by way of DNA. "We wish to suggest a structure for the salt of deoxyribonucleic acid (DNA). This structure has novel features which are of considerable biological interest," announced James Watson and Francis Crick, DNA's discoverers, with a historic understatement in 1953.

 As this famous quote indicates, DNA is a misnomer because it is technically a salt (sodium). Sodium is a critical human electrolyte and an excellent conductor of electromagnetism. Thus it is hardly surprising many researchers have determined, by way of empirical analysis, that DNA directly regulates the body's electromagnetics.

Through extensive kinesiological research, Leigh and I identified more than 3,000 energy signatures over the body-mind-spirit continuum of the human electromagnetic blueprint. That is probably just the tip of the iceberg. A traditional allergy or emotional clearing of this size would exceed, by far, the capacity of the healthiest nervous system. But when properly activated by sound combined with intention, *the superconductor that is DNA is designed to re-harmonize the entire bioenergy blueprint.*

When we speak of the body-mind-spirit continuum, we are not merely paying lip service to the interconnectedness of these elements. True healing means becoming "whole." Or as poet and metaphysical author Wynn Free eloquently puts it, healing is "that which removes ... the blockages from recognizing the existence of God within the self, and then becoming that Self."

From this perspective, assisting someone to achieve wholeness (as opposed to mere symptom remission or, worse, suppression) cannot be accomplished without helping that person address the entire range of genetic, physical, mental, emotional, spiritual and karmic energies that are in disharmony.

We are touching on a critical distinction between curing, which usually involves the patient giving away his or her power to an outside source, and healing, which cannot be done for but only by a person. In order to establish and maintain total wellbeing, it behooves us to reconnect with our birthright of inherent health and vitality and understand that we ourselves are responsible for our own healing.

We are touching on a critical distinction between *curing*, which usually involves the patient giving away his or her power to an outside source, typically a doctor trained in a limiting paradigm with respect to human potential, and *healing*, which in the final analysis cannot be done for but only *by* a person. In order to establish and maintain total wellbeing, it behooves us to reconnect with our birthright of inherent health and vitality and understand that we ourselves are responsible for our own healing.

One of my mentors was an important figure in the field of radionics, a chiropractor named David Tansley. Dr. Tansley, along with Alice Bailey and Helena Blavatsky, provided some of the foundation for my notions about the electromagnetic fields.

Following Tansley's lead, and supported by the quantum sciences' holographic view of the body, I began to understand the electromagnetic fields as our bioenergy template—an aspect of the hyperdimensional, "meta-genetic" blueprint that gives rise to and regulates the functions of our physical form. My theory was that when "mapped," these fields would reveal themselves as "ecosystems" where a number of interdependent factors work either harmoniously to create vitality or disharmoniously to produce disease.

The approach Leigh and I took to map the electromagnetic fields was relatively straightforward. Using kinesiology with ourselves and our clients, we began muscle testing to establish which elements (genetic, physical, mental, emotional, spiritual, etc.) were governed by which fields. We discovered an amazing poetic symmetry, a sacred geometry of almost breathtaking beauty in the way the fields are organized and work in concert.

This is the same sacred geometry that Horowitz references from an electrogenetic perspective and is intimately related to our molecular geometry that Merrill Garnett came to appreciate as music in his pioneering cancer research. "There is a harmony of the organism and a harmony in structure that allows the transfer of energy so that the organism can live and vibrate," writes Dr. Garnett. "Those harmonies and resonances recur and recreate the organism ... Ultimately, there is a musical or harmonic element within the organism ... This is molecular music, fragile, dependent, recurring under the right conditions, based in quantum echoes and hidden physics."

Sometimes, viewed from the perspective of disease, the musical and lyrical geometry of our bioenergy fields can appear very dark—recalling William Blake's "fearful symmetry"—but it is still poetry. For example, Leigh and I discovered that the third electromagnetic field is where cancer energies reside in many people.

I emphasize "many people" because one of our other discoveries was that, energetically speaking, all people are inherently and emphatically *not* the same. In fact, based on a wealth of kinesiological data, it appears there are twelve different "Electromagnetic Groups" of humans on this planet.

Energetically, these twelve groups correspond to the twelve pairs of cranial nerves, with each group contributing to humanity's "collective Mind." These twelve groups also align with the twelve acupuncture meridians, the twelve months, the twelve signs of the zodiac, Earth's twelve tectonic plates, and even the biblical Twelve Tribes. What unites these sometimes strikingly different energetic families is their shared "operating system": DNA.

Each Electromagnetic Group possesses a unique arrangement in its bioenergy blueprint that applies to all members.[5] This is truly an exciting revelation because in the context of DNA activation, it renders individual diagnosis unnecessary.

Etymologically, *diagnose* derives from Greek and can be thought of as to "read through" in order to achieve knowledge or *gnosis*. But as it too often is practiced in today's medicine, diagnosis tends to oversimplify complex processes while "locking in" a disorder in the sufferer's mind until it seems that nothing, or very little, can be done. This mentality helps explain the emergence in recent years of such deflating phrases as "disease management."

"A numbing, unquestioned acceptance of a given medical prognosis can stem from a variety of foundational beliefs," writes Barbara Marciniak, "yet it will all boil down to a strong underlying belief in personal powerlessness ... The largely ineffective, costly health-care system is sustained by such beliefs." The need for someone else to be "in charge of fixing and taking care of the body has created a cumbersome bureaucracy to deal with cradle-to-grave health concerns that are, for the most part, founded on conditioned fears contrived in the mind."

---

5 See Appendix C for an example.

Writes Richard Bartlett on this subject in *Matrix Energetics: The Science and Art of Transformation,*

> We don't get that much practice in the art of feeling good because our whole medical system is predicated on the treatment of disease, not wellness care. Medical practice is all about identifying symptoms, conditions, and treatment: seeing you as a person with problems. This process allows doctors to figure out which little reality box we can squeeze you into. And each time you're given a different diagnosis or explanation for why you have your symptoms, it further limits your awareness of what's possible. You're stuffed into smaller and smaller boxes, where you can do less and less about more and more. Pretty soon you feel restricted, despondent, and disempowered.

Dr. Bartlett adds that most people "have amazingly low expectations for what we think we can get back from the universe or out of life. Our experiences will conform very closely to the structure of our beliefs about life." All too often, in a sincere "effort to make progress, we end up going around in circles. Many forms of treatment or therapy seem to reinforce the problem mind-set, despite good intentions."

The obsessive focus on labeling in the Western medical paradigm, combined with patients content to give away their power by consenting to being labeled, is arguably a main reason iatrogenic or doctor-induced deaths have become a crisis not only in the United States, but in many parts of the world—with thousands of people dying unnecessarily under medical supervision every year. By emphasizing the importance of the individual's commitment to conscious personal mastery (see Glossary) as a prerequisite for becoming whole, the Regenetics

Method represents a purposeful shift away from the diagnostic model.

> **By emphasizing the importance of the individual's commitment to conscious personal mastery as a prerequisite for becoming whole, the Regenetics Method represents a purposeful shift away from the diagnostic model.**

It is worth adding that Leigh and I also perform Regenetics sessions with nonattachment, intending only the client's highest good, since (as mentioned in Chapter One) studies have shown that non-directed prayer is statistically more effective in healing than prayer with an agenda. To "potentiate" a person, we simply use surrogate muscle testing to determine the Electromagnetic Group, whose Schematic is used primarily by the client following the session for self-educational purposes, and perform the same generalized DNA activation that is good for all Electromagnetic Groups.

But to return to the example of the third electromagnetic field. In addition to cancer energies, this field is also where we often find radiation energies. It is common knowledge there exists a direct link between cancer and radiation. Now, consider the emotional content of the third field. The primary emotions that exist in the energetic ecosystem with cancer and radiation are fear and related feelings of anxiety, worry, panic, and terror.

Now, guess which toxins show up in the third field. Chemicals. Pharmaceuticals. Drugs. Cigarettes. They are just sitting there in the third electromagnetic field contributing to cancer with radiation and fear-based emotions.

This cannot be mere coincidence. We live in a global culture that in some aspects could be described as "Orwellian," take handfuls of toxic drugs for a headache, ingest aspartame on a daily basis, are bombarded continuously by radiation from computers, cellular telephones and microwave ovens, and have an alarmingly high incidence of cancer.

It took months and a lot of vials to map all the major energies in the body's electromagnetic fields. Leigh and I spent six months developing our ideas in South America, where we performed the first Potentiation Electromagnetic Repatterning on ourselves that restored my physical wellbeing and took care of Leigh's asthma and environmental allergies. Then we began offering Potentiation to others, many of whom have reported remarkable results as evidenced by the wide-ranging Testimonials provided in Appendix A.

In order to complete our work on Potentiation, however, we first had to set aside what we had been taught about DNA (that it is merely a biochemical protein-assembly code) and understand DNA's vitally reciprocal relationship with the body's electromagnetics. Only then were we in a position to explore avenues for stimulating the human genome's extraordinary self-healing potential.

5
Resetting the Bioenergy Blueprint via DNA

After mapping the electromagnetic fields, Leigh and I realized we had to find a way to press the "reset button" on this complex bioenergy blueprint. Coming from my NAET/BioSET perspective, at first I thought we had to develop a technique to "clear" all the energies that somehow were "blocked." Going back to the example of the third electromagnetic field discussed in the previous chapter, I assumed that in order to begin addressing an entrenched condition like cancer, one somehow had to remove the energetic "roadblock" formed by radiation, fear-based emotions, pharmaceuticals, cigarettes, etc.

It was at this stage I began to understand that the nervous system was never meant to repattern the human bioenergy blueprint; that only DNA, our biocomputer, can build a new energy body; and that therefore, some other method of initiating electromagnetic repatterning besides acupressure stimulation of the meridian system had to be found.

We went to DNA because it was the obvious choice. DNA contains our genetic code, the master blueprint for our biology. It literally creates us through a protein-assembly process known as transcription. In an article reprinted in *DNA Monthly*, Dr. Stephen Linsteadt offers the following excellent summary of genetic transcription that takes into account both the biochemical and electromagnetic aspects of this life-creating process:

The cell's innermost center is composed of ribonucleic acid and proteins (all molecules). The antenna or filament strand-like configuration of DNA allows the molecules to receive and transmit electromagnetic frequency information along its nucleotide bases, creating resonance reactions in genetic nucleotide triplets that create the template for the formation of messenger RNA (mRNA). Once mRNA has formed, it leaves the cell nucleus and attaches to structures known as ribosomes. Using raw material from cells, ribosomes produce proteins by following the sequence as instructed by mRNA. Proteins, in turn, go about their jobs inside or outside cells based on the original instructions passed down from the electromagnetic coding from DNA to RNA and finally to ribosomes. This process is known as *transcription* and provides the means for electromagnetic frequency oscillations, the body's master conductor, to interact with the cell's command center to instruct what notes to play, when, how loud, how long, etc., in order to maintain the precision and harmony of the whole body's vibratory and cellular orchestra.

To transcribe can be defined as to copy in writing, to produce in written form, or to arrange music for a different instrument. In other words, as the above quote suggests, we come into being, at our molecular level, through a process with striking affinities to *composition*.

It is extremely interesting to consider the privileged place of song, storytelling and words in creation myths. Anyone who has undertaken a comparative study of religions probably has been struck by the universal role of sound and language in such myths. Genesis 1:3 relates, "And God *said*, Let there be light: and there was light" (my emphasis). In the New Testament, John states, "In the beginning was the Word," an idea paralleled in the Vedas where we read, "In the beginning was Brahman with whom was the Word."

The *Popol Vuh* from the Mayan tradition insists that the first humans were brought into existence by speech, just as the ancient Egyptians believed that the god-men Thoth and Ra created life through language. Writes Joachim-Ernst Berendt in a masterful study of sound and creation, *The World Is Sound—Nada Brahma*,

> In Egypt, the "singing sun" created the world with its "cry of light." In an ancient Egyptian scripture it is written that "through the tongue of the Creator ... all Gods and everything in existence were born ... Atum and everything divine manifest themselves in the thought of the heart and in the sound of the tongue." The symbol for "tongue" in Egyptian hieroglyphics can also mean "word"; it is the tongue that forms the sound that in turn carries the word.

Explains Berendt, "In the beginning was the sound, the sound as *logos*. If you remember, God's command 'Let there be ...' at the beginning of the biblical story of creation was first tone and sound. For the Sufis, the mystics of Islam, this is the core of things: God created the world from sound."

"In India," Berendt continues in his multicultural examination of this all-important theme, "in the *Aitereya-Upanishad* ... [o]f Brahma it is said: 'He meditated a hundred thousand years, and the result of his meditation was the creation of sound and music' [...] Thus, the first act of creation was the creation of sound. Everything else came after and through it."

Consistent with this universal speech-based cosmology, the healing tradition immortalized in the Bock Saga originating in Finland is based on memorization and utterance of sacred sounds. This Saga, which Horowitz describes at length in *DNA: Pirates of the Sacred Spiral,* is an elaboration of a time-honored oral technology employing sound and light based on a "spiritual understanding of how to work with 'nature orally'"—or "naturally."

"Here, in ancient mythology," writes Horowitz, "is the relationship between genesis, genetics, and the spoken word. Also implied is the concept of wholistic health hinging on oral functions." Horowitz points out that today's neurophysiologists have determined that fully "one-third of the sensory-motor cortex of the brain is devoted to the tongue, oral cavity, the lips, and speech. In other words, oral frequency emissions (i.e., bioacoustic tones) spoken, or sung, exert powerful control over life, vibrating genes that influence total well-being and even evolution of the species."

Since the start of the Human Genome Project and the chromosomal mapping of the human genetic structure, there has been a tendency even in mainstream science to regard DNA as the alphabet through which, essentially, we are *written* into existence.

Another metaphor often helpful in visualizing the somewhat complicated mechanism of genetic composition derives from music. The building of our protein structures, of our cells that form our tissues and organs, starts with RNA transcription of specific codes contained in DNA. We can imagine RNA as a magnetic recording tape that "plays" data stored in DNA as a composition of "notes" in a configuration of amino acids that create bars of "music" called proteins.

Leigh and I realized that if we were to activate what we saw as an extraordinary latent potential in DNA, one perhaps capable of transforming both consciousness and physiology we intuited along with a growing number of scientists including Linsteadt, Horowitz, Gregg Braden and Bruce Lipton, we had to find or develop a way to access DNA without laboratories or test tubes. But how do you do that? How do you activate DNA without physically manipulating it?

At this stage, we were fortunate enough to be given a copy of *The Cosmic Serpent: DNA and the Origins of Knowledge* by French anthropologist Jeremy Narby. Dr. Narby spent years studying the healing techniques of shamans (medicine men) in the Amazon.

His account, anthropologically as well as scientifically, is riveting and was particularly helpful in developing the Regenetics Method. In one telling passage, Narby writes, "DNA is not merely an informational molecule, but ... also a form of text and therefore ... is best understood by analytical ways of thinking commonly applied to other forms of text. For example, books."

Coming from my background in literary theory and fiction writing, this way of looking at DNA as a book was extremely appealing. More than anything, it just made sense. Narby clearly is saying we can learn to read DNA. By implication, he is suggesting we also can learn to write, or rewrite, the genetic code. This is how I can speak, in all seriousness, of "textual healing."

An alternative way to conceptualize what I am calling "rewriting" is to imagine that DNA contains a subtext resembling a series of footnotes that can be scrolled up onscreen. In this scenario, no rewriting or reprogramming is required. The program for our new and improved energy body already exists in what mainstream science has dismissed as "junk" DNA.

Most classically trained geneticists have admitted they have virtually no idea why over ninety percent of our DNA even exists. This is especially provocative given that over ninety percent of our brain also is unused.

Narby clearly is saying we can learn to read DNA. By implication, he is suggesting we also can learn to write, or rewrite, the genetic code.

Most of DNA appears to be nonsense. A lot of it is in the form of palindromes, puzzling sentences that read the same forward and backward. "Junk" DNA consists primarily of "introns," considered noncoding genetic sequences, as opposed to "exons" that have an identifiable coding function in building our protein structures through RNA transcription. In other words, as shown in Figure 2, exons clearly do something, while introns supposedly do not.

Fortunately, more and more scientists who have asked how nature could be so inefficient are beginning to rethink this dogma that ultimately raises more questions than it answers. Recent research has shed light on intense epigenetic as well as meta-genetic activity in "junk" DNA, which appears to have much more to do with creating a specific species than previously thought. For example, if we only look at the small portion of DNA composed of exons, perhaps you have heard there is little difference, genetically speaking, between a human being and a fruit fly? There is also practically nothing at the level of exons that distinguishes one human being from another.

Others who have studied the mystery of "junk" DNA insist the as little as three percent of the human genome directly responsible for protein transcription does not contain enough information to build *any* kind of body. Faced with

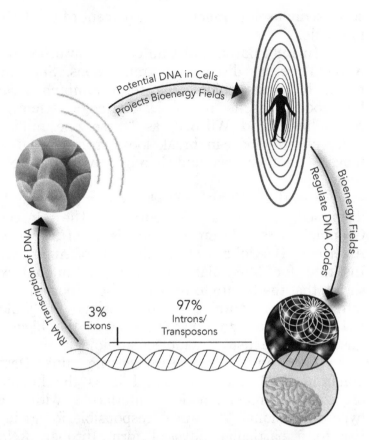

Figure 2: The Ener-genetic Composition Process
The above diagram illustrates how body building is both genetic, involving RNA transcription of DNA codes to create cells, and energetic, dependent on the interface between the electromagnetic fields and "junk" or potential DNA for regulation of cellular composition. This diagram also shows how potential DNA's transposons can be prompted directly by consciousness, internal (personal) and external (universal), to modify cellular replication.

this mystifying scenario, more and more scientists are paying attention to curious structures called "jumping DNA" or "transposons" found in the supposedly useless ninety-seven percent of the DNA molecule.

In 1983 Barbara McClintock was awarded the Nobel Prize for discovering transposons. She and fellow biologists coined the term *jumping DNA* for good reason, notes leading-edge scientific researcher David Wilcock, as "these one million different proteins can break loose from one area, move to another area, and thereby rewrite the DNA code."[6]

Clearly, "junk" DNA was prematurely dismissed. In an article entitled "The Unseen Genome" in the November 2003 issue of *Scientific American*, Dr. John Mattick, director of Australia's Institute for Molecular Bioscience, is quoted as saying that the failure to recognize the importance of introns (to say nothing of transposons) in "junk" DNA "may well go down as one of the biggest mistakes in the history of molecular biology."

Leigh and I propose we rename "junk" DNA *potential DNA* and understand it as the human organism's meta-genetic interface with a hyperdimensional "life-wave" responsible for giving rise to a particular physical form through RNA transcription of DNA codes. We will return to this critically important idea a bit later.

DNA, whether coding or noncoding, whether exons, introns or transposons, is composed of an "alphabet" of four basic "letters" for creating nucleotides that combine to form sixty-four different

---

[6] Technically, transposons that move to a new location using RNA are called "retrotransposons," of which there are three principal types—SINEs, LINEs, and HERVs—potentially involved in large-scale genetic transformation (Kelleher). For the sake of simplicity, I will employ the term *transposon* to indicate all such structures capable of jumping from one chromosomal position to another.

"words" used to build a virtually limitless number of "sentences" called genes. The number 64 is especially interesting given another of our inspirations, J. J. Hurtak's *The Keys of Enoch*. *The Keys of Enoch* is an elaboration of the "keys" for creating a higher energy body. Significantly, there are sixty-four keys, just as there are sixty-four nucleotide combinations. Hurtak obviously is writing about actualizing a transformational potential in our genetics.

> *Leigh and I propose we rename "junk" DNA potential DNA and understand it as the human organism's meta-genetic interface with a hyperdimensional "life-wave" responsible for giving rise to a particular physical form through RNA transcription of DNA codes.*

I realize this is a lot of information, especially for the lay reader, for whom I have endeavored to simplify this material as much as possible. As these concepts are among some of the most life-changing I have encountered personally in my research, I encourage you to go over this chapter and any others that seem complicated as many times as necessary until the ideas sink in.

To summarize to this point, it is accurate to say that DNA is a form of text with its own alphabet, and that we can learn to read as well as rewrite DNA, in the process activating the genetic program designed to turn our introns into exons (via transposons) and create new protein transcription sequences that lead to regeneration, or *re-gene-ration*.

6
Sound, Intention & Genetic Healing

In a very intriguing section of *The Cosmic Serpent*, Narby includes snippets from his personal journals. One entry is of particular interest at this point in our discussion of DNA:

> According to shamans of the entire world, one established communication with spirits via music. For [shamans] it is almost inconceivable to enter the world of spirits and remain silent. Angelica Gebhart Sayer discusses the visual music projected by the spirits in front of the shaman's eyes. It is made up of three-dimensional images that coalesce into sound, and that the shaman imitates by emitting corresponding melodies.

In a provocative footnote to himself, Narby adds, "I should check whether DNA emits sound or not."

One school of thought insists that humans actually are made of sound and that DNA itself may be a form of sound. Drawing on meticulously documented research, Horowitz explains that DNA emits and receives both phonons and photons, or waves of sound and light.

In the 1990s, according to Horowitz, "three Nobel laureates in medicine advanced research that revealed the primary function of DNA lies not in protein synthesis ... but in electromagnetic energy reception and transmission. Less than three percent of DNA's function involves protein manufacture; more than ninety percent functions in the realm of bioacoustic and bioelectric signaling." In recent years, a fascinating artistic field called DNA music has begun to flourish. It therefore seems appropriate, at the very least, to compare DNA to a keyboard with a number of keys that produce the music of life.

But what if on some level we *are* made of sound? What if in the beginning *was* the Word? What if the music of the spheres is no myth? What if we ourselves are a harmonic convergence? What if the holographic grid of our being is a linguistic and musical interface between higher-dimensional light, which might be considered a form of divine thought or *intention*, and *sound* in higher-dimensional octaves? After all, String theory, with offshoots that still enjoy some popularity in the scientific community, posited the existence of many different, theoretically accessible dimensions that appear notationally linked much like strings on a guitar.

Narby repeatedly makes the point that shamans use sound because this allows them to transform some aspect of the genetic code. If DNA is indeed a text, a keyboard, a musical score; if it is true this score can be rewritten so that it plays a new type of music; and if we live not just in a holographic but in a *harmonic* universe, then it seems entirely plausible, returning to an earlier idea, that *our electromagnetic fields are concentric spheres of hyperdimensional sound.*

When Leigh and I began developing the Regenetics Method, we discovered through muscle testing that each of the nine electromagnetic fields corresponds not just to a chakra, but to one of nine third-dimensional sound octaves. Energetically, our research indicates that at our present evolutionary stage, humans are built of a vertical series of nine light-processing chakras interfacing with nine concentric electromagnetic fields (which are sonic in nature) to form the three-dimensional holographic matrix that produces our physical body. Through

DNA activation, as detailed in the next chapter, this bioenergy blueprint can be upgraded to an "infinity circuit" based on eight chakras and eight fields (Figures 1 and 3).

Moreover, it appears from ongoing research that at the genetic level, sound gives rise to light—a microcosmic assertion consistent with the basic macrocosmology underwriting the Regenetics Method elaborated in subsequent chapters.

In a paper entitled "Quantum Bioholography" appearing in *The Journal of Non-Locality and Remote Mental Interactions,* a team of researchers headlined by Richard Alan Miller outline a compelling model of ener-genetic composition resulting in "precipitated reality": "Superposed coherent waves of different types in the cells interact to form diffraction patterns, firstly in the acoustic [sound] domain, secondly in the electromagnetic [light] domain." This leads to the manifestation of physical form as a "quantum hologram—a translation between acoustical and optical holograms." Significantly, this sound-light translation mechanism that creates the somatic experience of reality functions in the genome.

This is not the place to provide a full treatment of the impressive science of quantum bioholography. Rather, I wish to emphasize that according to this model that is attracting many proponents as more and more of its precepts are confirmed, it is becoming apparent that DNA directs cellular metabolism and replication not just biochemically, but electromagnetically through a chromosomal mechanism that translates sound into light waves, and vice versa.

According to the holographic model of genetic expression, sound and light, or phonons and photons, establish a sophisticated communication network throughout the physical organism that extends into the bioenergy fields and back to the cellular and subcellular levels (Figure 2).

Recalling Edgar Cayce's prediction that "sound would be the medicine of the future," Jonathan Goldman in *Healing Sounds* coined the following inspirational formula: sound + intention = healing. If we define intention as a form of conscious light energy equivalent to thought, an idea consistent with many shamanic traditions such as that of the Toltecs of Mesoamerica, we can translate Goldman's formula as:

SOUND + LIGHT = HEALING.

Recently, the ability of sound and light to heal DNA was documented scientifically by a Russian research team of geneticists and linguists. Russian linguists studying a leading-edge branch of semiotics known as "Genetic Linguistics" discovered that the genetic code, particularly that in potential DNA, follows uniform grammar and usage rules virtually identical to those of language. This revolutionary research grew out of Jeffrey Delrow's stunning discovery in 1990 that *the four nucleotide bases of DNA inherently form fractal structures closely related to human speech patterns.*

As explained by Russian biophysicist Peter Gariaev, the developer of wave-genetics who spearheaded this genetic-linguistic research, "a group of scientists headed by M. U. Maslov and myself developed a theory of fractal representation of natural (human) and genetic languages. Among other things, this theory postulates that the 'quasi-speech' of DNA possesses a potentially inexhaustible supply of 'words' and, moreover, that 'texts,' 'phrases' and 'sentences' in DNA transform into the letters and words used in human speech."

At a stroke, this evidence eviscerates many modern linguistic theories by proving that language did not appear randomly, but grows organically out of humanity's shared genetics. On a related note, human language, as pointed out by well-known linguist Noam Chomsky, clearly does not exhibit a linear development from animal to human speech patterns. "This poses a problem for the biologist," concedes Chomsky, "since, if true, it is an example of a true 'emergence.'"

In *The God Code*, Gregg Braden further investigates this notion of genetic-linguistic "emergence" of biological species by showing that the ancient four-letter Hebrew name for God (YHVH, the Tetragrammaton) is actually code for DNA based on the latter's chemical composition of nitrogen, oxygen, hydrogen, and carbon. This assertion, with its vast implications relative to DNA's universal role as a divine language spoken through the body, has been peer-reviewed and accepted by many scholars of Hebrew.

According to author Zecharia Sitchin as well, human DNA is deeply linguistic in nature. In *The Cosmic Code*, Sitchin speculates,

> Could the reason for the three-letter Hebrew root words be the three-letter DNA language—the very source, as we have concluded, of the alphabet itself? If so, then the three-letter root words corroborate this conclusion ... 'Death and life are in the language,' the Bible states in Proverbs (18:21). The statement has been treated allegorically. It is time, perhaps, to take it literally: the language of the Hebrew Bible and the DNA genetic code of life (and death) are but two sides of the same coin ... The mysteries that are encoded therein are vaster

than one can imagine; they include among other wondrous discoveries the secrets of healing.

Fritz Albert Popp's Nobel Prize-winning research establishes that every cell in the body receives, stores and emits coherent light in the form of biophotons. In tandem with biophonons, biophotons maintain electromagnetic frequency patterns in all living organisms.

In the words of Stephen Linsteadt, this matrix that is produced and sustained by frequency oscillations "provides the energetic switchboarding behind every cellular function, including DNA/RNA messengering. Cell membranes scan and convert signals into electromagnetic events as proteins in the cell's bi-layer change shape to vibrations of specific resonant frequencies." Emphasizing that every "biochemical reaction is preceded by an electromagnetic signal," Linsteadt concludes, "Cells communicate both electromagnetically and chemically and create biochemical pathways that interconnect all functions of the body."

Dr. Gariaev and fellow Russian biophysicist Vladimir Poponin also have studied DNA's extraordinary electromagnetic properties. Their research reveals that DNA has a special ability to attract photons, causing the latter to spiral along the helix-shaped DNA molecule instead of proceeding along a linear path. In other words, DNA has the amazing ability—unlike any other molecule known to exist—to bend or weave light around itself.

In addition, it appears that a previously undetected form of intelligent light or intention energy (emanating from higher and/or parallel dimensions and distinguishable from both gravity and electromagnetic radiation) which Dr. Eli Cartan first termed "torsion" in 1913 after its twisting movement through the fabric of space-time, gives rise to DNA.

Many decades later, the concept of torsion energy was still alive and well enough to inspire an entire generation of Russian scientists, who authored thousands of papers on the subject in the 1990s alone. "A unified subliminal field of potentially universal consciousness apparently exists," writes Horowitz on the subject of the Russian studies, "and may be explained as emerging from a previously overlooked physical vacuum or energy matrix."

Actually, the ancient Greeks were well aware of this potent energy, calling it "aether" and understanding that it is directly responsible for universal manifestation. In the 1950s, Russian scientist Nicolai Kozyrev conclusively proved the existence of this life-giving subspace energy, demonstrating that, like time, it flows in a sacred geometric spiral (inherently fractal in nature, like both DNA and human language) resembling the involutions of a conch shell that has been called *Phi*, the Golden Mean, and the Fibonacci sequence.

In the face of overwhelming evidence of its existence, Western scientists are returning to the notion of aether using such phrases as "zero point energy" and "vacuum potential."[7] Not too long ago,

---

[7] A new model called the Electric Universe theory is beginning to challenge the notion of zero point energy and the like by suggesting that such "missing matter," rather than being "multidimensional," actually exists in an electrical, or plasma, state in our own dimension. If true, the Electric Universe theory potentially could do away with the Big Bang by demonstrating that the universe is continuously self-generating and thus has no beginning. To my knowledge, the relationship between matter in an electrified plasma state and torsion energy has yet to be described coherently. Theoretically as well as practically, the two could be intimately related. See Footnote 22 in

physicists Richard Feynman and John Wheeler went so far as to calculate that the amount of torsion energy contained in a light bulb literally could bring the world's oceans to a boil!

> *This breakthrough research in the temporal physics of subspace establishes that torsion energy permeates the entire multidimensional galaxy and not only is responsive to, but actually may* be, *consciousness creatively experiencing itself in time.*

This breakthrough research in the temporal physics of subspace—part of an emerging scientific field known as *temporology*—establishes that torsion energy permeates the entire multidimensional galaxy and not only is responsive to, but actually may *be*, consciousness creatively experiencing itself in time.

"To put it as bluntly as possible," writes David Wilcock, "you cannot separate consciousness and torsion waves—they are the same thing. When we use our minds to think, we are creating movements of electrical impulses in the brain, and when any electrical energy moves, torsion waves are also created."

In his popular blog, Wilcock expounds on the subject of torsion energy and its relationship to DNA: "For years now we've been saying DNA is a wave—spiraling 'nonliving' material together into a molecule. Now there's proof! A huge validation has just been handed to us from the annals of mainstream science [...] Loose inorganic materials can spontaneously and intelligently spiral together to form DNA ... in the cold emptiness of space! One DNA molecule is as complex as an entire encyclopedia—so without a higher, organized

---

Chapter Fourteen.

intelligence guiding the process, we really cannot explain this."

The research Wilcock references involved controlled experiments by Dr. Ignacio Ochoa Pacheco in which DNA formed out of a hermetically sealed container—containing only distilled water and sand heated to the point of killing any living organisms therein—when exposed to consciousness, or torsion waves. Wilcock observes that Pacheco's research "validates the work of my colleague Dr. Dan Burisch ... who has taken the process even further and observed what appear to be micro-wormhole structures emerging from 'the vacuum.' They seem to act as precursors to primitive cellular structures developing spontaneously."

If this evidence seems fantastical, consider the article "Dust 'Comes Alive' in Space" appearing in the *UK Times Online* in August of 2007, where it was reported that "an international panel from the Russian Academy of Sciences, the Max Planck institute in Germany and the University of Sydney found that galactic dust could form spontaneously into ... double helixes [...] and that [these] inorganic creations had memory and ... power to reproduce themselves [...] The particles are held together by electromagnetic forces that the scientists say could contain a code comparable to the genetic information held in organic matter."

Offering the simplest explanation of these interrelated phenomena involving the spontaneous appearance of DNA and DNA-like formations, Wilcock writes that "DNA is a physical materialization of what torsion-waves look like at the tiniest level. Don't forget we are dealing with intelligent energy ... This, of course, strongly suggests that life could form spontaneously from inert 'nonliving' material." In Chapter Nine, we will examine how the tendency to form DNA-shaped helices can be observed even at the level of whole nebulae, strongly suggesting that the spiral shape commonly observed in biological organisms is, in fact, a universal form emerging from a cosmic background field of torsion radiation.

According to the Russian findings, notes Wynn Free, "this spiraling 'torsion' energy could actually be the substance of our human souls, and is therefore the precursor to the DNA molecule ... It already exists in the fabric of space and time before any physical life emerges." Elsewhere, Free remarks of transposons that these tiny segments of DNA can travel along the genome activating different parts of it when prompted by consciousness.

In keeping with Dr. Gariaev's "Wave-based Genome" theory, Free concludes that DNA functions "somewhat like a computer chip, with different sections that can either be 'on' or 'off.'" Thus we easily can imagine how the torsion waves of human consciousness could program, or reprogram, DNA's binary code (Figure 2).

Behind the DNA molecule exists a template of consciousness that directs the formation of organisms at the level of DNA. Modify this subtle energy blueprint, alter this consciousness through meta-genetic means, and we potentially transform organic expression.

Beyond any reasonable doubt, based on overwhelming scientific evidence, behind the DNA molecule exists a template of consciousness that directs the formation of organisms at the level of DNA. Modify this subtle energy blueprint, alter this consciousness through meta-genetic means, and we potentially transform organic expression.

The Gariaev group has demonstrated that chromosomes function much like (re)programmable holographic biocomputers employing DNA's own electromagnetic radiation. Their research unambiguously states that human DNA is composed literally of genetic "texts"; that chromosomes both produce and receive the information contained in these texts in order to encode and decode them; and that chromosomes assemble themselves into a holographic grating or lattice designed to generate and interpret highly stable waves (from frequency oscillations in DNA) of sound (carried by radio signals) and light that direct all biological functions.

In other words, explain longtime genetics researchers Iona Miller and Richard Miller in an article reprinted in *DNA Monthly* based partly on Gariaev's findings entitled "From Helix to Hologram," the genetic "code is transformed into physical matter, guided by light and sound signals."

One revolutionary implication (of many) of this research is that, to activate DNA and stimulate healing at the cellular level, we simply can use our species' supreme expression of creative consciousness: words.

Decades of research by Japanese scientist Kikuo Chishima substantiate this assertion. Dr. Chishima's work strongly suggests that red blood cells are formed not in bone marrow, as is commonly believed, but in the intestinal villi. Red blood cells appear to be 1) guided by genetic frequency oscillations (ultimately governed by the bioenergy blueprint) and 2) capable of synthesizing DNA in order to differentiate into specific types of cells, which then migrate via the 90,000-mile-long capillary system to wherever they are needed. Writes Linsteadt, "This open-ended system that connects to the lymphatic system, the meridian system and the connective tissue provides communication pathways for the flow of information and cellular instructions from the electromagnetic energy matrix."

One revolutionary implication (of many) of this research is that, to activate DNA and stimulate healing at the cellular level, we simply can use our species' supreme expression of creative consciousness: words. While Western researchers clumsily cut and splice genes, Gariaev's team developed sophisticated devices (quantum biocomputers) capable of influencing cellular metabolism and stimulating tissue regeneration through sound and light waves keyed to human language frequencies.

Using this method, Gariaev proved that chromosomes damaged by X-rays, for instance, can be repaired; and even more strikingly, that a diseased pancreas in rats can be *regrown*. Moreover, this was accomplished *noninvasively* by merely applying vibration and language, or sound combined with intention, or *words*, to DNA.

According to Iona Miller and Richard Miller, "Life is fundamentally electromagnetic rather than chemical, the DNA blueprint functioning as a biohologram which serves as a guiding matrix for organizing physical form." Possibly the most far-reaching implication of the research cited in this chapter is that DNA can be activated through conscious linguistic expression (somewhat like an antenna) to reset the bioenergy fields, which in turn (like orbiting communication satellites) can transmit radio and light signals to restore proper cellular structure and functioning of the human body.

7
Sealing the Fragmentary Body

Those with highly evolved consciousness such as spiritual teachers always have insisted that the human body is genetically (re)programmable by words in the form of songs, poems, prayers, affirmations, or mantras. The words must be harmonically attuned to the organism and the intention behind them impeccable.

This is why although DNA activation has become trendy, results can vary enormously. The more advanced the individual healer's consciousness, and the more s/he is ener-genetically balanced, the less need there is for machines. It has been claimed that, in theory, one can achieve life-transforming results in wave-genetics unassisted by external technologies (Mohr). It is my personal belief that the individual can achieve far superior results than is possible with machines, especially where not just physical healing but growth in consciousness is desired.

Citing a variety of scientific studies that prove sound can change human brainwaves as well as heartbeat and respiration, Goldman highlights the developments in the field of sound therapy credited to such medical pioneers as Dr. John Diamond, Dr. Peter Manners and Barbara Hero, all of whom have designed mechanical instruments for healing through sound. Clearly, however, Goldman believes the human voice is the ultimate healing instrument.

Some shamanic healers go so far as to insist that the transformative power of the human voice cannot be reproduced digitally and retain its full character—that the digital recording is comparable to a clone, "possessing form but lacking spirit"—which has led me to question the effectiveness of DNA activation CDs and similar technologies.

Allow me to direct your attention to the notion, found in so many religions and mythologies, of a "fall from grace" that created a rift in the universe, a disruptive force that engendered duality and the experience of separation.

In Christianity this often is termed "original sin." In one Hindu myth, human consciousness began as a tiny ripple that chose to leave the ocean of cosmic consciousness. As it awoke to itself, our consciousness forgot it was part of the infinite cosmic ocean and found itself washed ashore and imprisoned in a state of isolation. Wilcock calls this perceived separation from Source the "Original Wound" and wisely remarks that it is "the basis behind all suffering, and also ... the final key to enlightenment."

Science has its own versions of the fundamental duality at the heart of existence as we experience it. The particle-wave duality, in which atomic components are simultaneously particles and waves, is a primary example. Not surprisingly, DNA also has been shown to possess a version of the particle-wave binarism. "In accordance with this duality," writes Horowitz, "DNA codes all living organisms in two ways, both with the assistance of DNA matter involving RNA and enzymes for protein synthesis, and by DNA sign wave functions, including coding at its own laser radiation level that functions bioholographically" (Figure 2).

From the outset, the holographic model has focused on the duality inherent in human experience. Dr. Karl Pribram first theorized a neural hologram in the brain's cerebral cortex operating in tandem with a subatomic or universal hologram—a micro-macrocosmic, fractal interface summed up by

Horowitz when he states that "a hologram within a hologram produces life as a function of creative consciousness."

In *Wholeness and the Implicate Order*, David Bohm also describes the brain as a hologram designed to interpret a larger hologram—the cosmos. "In this dualistic holographic model," explains Horowitz, "inseparable interconnectedness of holographs, including that of the Creator with the created, underlies human existence." Human existence, in turn, to quote Iona Miller and Richard Miller, is rooted in genes serving as "holographic memories of the existential blueprint."

As mentioned earlier, through kinesiological testing Leigh and I discovered and mapped a total of nine electromagnetic fields in humans. I offer that the initial blueprint for our creation, however, was one in which instead of nine, we had only *eight* fields corresponding to *eight* chakras.

This is a pivotal concept for anyone interested in genuine, permanent healing. There are many reasons why I insist that our true bioenergy blueprint is based on the number 8. The one I offer now is of a visual nature. What do you get when you turn the number 8 on its side? An infinity sign. This is our infinite nature, our divine birthright expressed in a symbol.

Perhaps you are familiar with the theosophical teachings of Bailey and Blavatsky or Tansley's writings on radionics. All three present a model of the human bioenergy template with only seven fields. Vedic teachings also are based on seven energy centers. But kinesiologically, at this stage of human development there are clearly nine fields, not counting a tenth we call the Source or Master Field that corresponds in astrophysical terms to Galactic Center (as covered in detail in Part II) and to *Nezah*

Figures 3a & 3b: Sealing the Fragmentary Body
The first image (Figure 3a) shows a typical human
bioenergy blueprint with nine electromagnetic
fields/*chakras* and a Fragmentary Body, envisioned as
an energetic disruption in the second field/chakra
from the bottom. The second image (Figure 3b) shows
a "potentiated" bioenergy blueprint with an "infinity
circuit" of eight fields/chakras. Note how sealing the
Fragmentary Body replaces fragmentation and duality
with harmony and sacred geometry, allowing for the
free flow of bioenergy throughout the body.

Figure 3b.

or Eternity in the kabalistic Tree of Life.

Bailey, Blavatsky and Tansley were right, however, when it comes to the second electromagnetic field. This field (with the corresponding "sex" chakra) has been called the *Fragmentary Body* (Figure 3a).

When mentioned in the esoteric literature, the Fragmentary Body is considered highly problematic. This is because the second electromagnetic field resonates as a "Frankenstein's monster" of energies that simply do not add up, that in many cases do not even appear to belong in the human body. For example, energies associated with all types of parasites attach to the second field.[8]

In every other electromagnetic field that governs a population of microorganisms, many of these are beneficial and undoubtedly belong in the body. For instance, in the seventh field of the Electromagnetic Group whose structure is shown in Appendix C, we find intestinal flora, which play a crucial role in creating a healthy biological terrain. But in the second field, we find only parasites, which—far from contributing to health—siphon off the host's life energy.

Each electromagnetic field also governs specific organ systems. The two organ systems found in the second field of all twelve Electromagnetic Groups are the reproductive system and the mouth: our (pro)creative systems. The intimate relationship between these seemingly distinct systems appears in the way we conceptualize and describe creativity. Authors "give birth" to a novel, "conceive" an idea, just as a poetic organ called the uterus "utters" a fetus into the world.

---

[8] It has been asked occasionally of Leigh and myself how we "discovered" the Fragmentary Body. To which we respond that, in a sense, we discovered nothing. Some type of "problem" with the second chakra often is intimated in various metaphysical traditions. Our contribution has been an elaboration (at the level of scientific and kinesiological data) of something intuitively sensed for centuries. As to why this energy center is not depicted as a fragmentation in holy texts, we suggest that just as with the Catholic Church, much was hidden for those with eyes to see and ears to hear in mystical traditions throughout the world, including that of the Vedics.

Developing the Regenetics Method led Leigh and myself overwhelmingly to a cosmology with a creation scenario where something disruptive occurred. This is not a judgment, simply an observation.

In the beginning was literally the Word, and something divisive resulted. Somebody spoke and birthed a dualistic universe of opposites, one with a Great Rift running through the middle mirrored overhead in the Milky Way.[9] In the microcosm of our energy body, in keeping with the ancient dictum "As above, so below," this Great Rift or Original Wound is paralleled in the second electromagnetic field.

We can envision the Fragmentary Body as an energetic vacuum that to a large degree separates spirit and matter by keeping hyperdimensional torsion energy from filling our electrogenetic matrix until we become "enlightened" in the flesh.

We can envision the Fragmentary Body as an energetic vacuum that to a large degree separates spirit and matter by keeping hyperdimensional torsion energy from filling our electrogenetic matrix until we become "enlightened" in the flesh.

The word *enlighten* literally means to light up, to illuminate. The Fragmentary Body is an anti-enlightenment consciousness vacuum, a systemic

---

[9] On this subject, Judith Bluestone Polich writes, "The 2012 alignment occurs when the December solstice sun conjuncts the crossing point of the Milky Way in Sagittarius. An area in the sky called the dark rift—known to the Maya as the Xibalba bi, the road to the underworld—points right to this crossing point. The crossing point is found at the center of our galaxy, and the Maya called it the sacred tree. To them it indicated the place of creation." See Part II.

bioenergy drain that, until "sealed," limits our ability to embody the light of higher consciousness. When properly sealed through DNA activation, however, this field that once represented an energetic liability becomes the locus for the human being's healing into a consciousness and physiology capable of expressing divine radiance.

In recent years, physicists have begun to acknowledge the existence of particles called tachyons, a powerful form of energy popularized by *Star Trek*. But tachyons are not science fiction. Tachyons are particle-waves belonging to the lepton family that, according to the evidence, fail to obey the law of gravity. In other words, they appear to travel faster than light.

Even normal quantum particles such as electrons have been shown to communicate "telepathically" with each other at a distance, as if connected at an aetheric level. As previously remarked, some members of the scientific community have gone so far as to resurrect the Greek term *aether* for the torsion-wave medium that is "empty" space. A variety of other scientific phrases, including "dark energy" and "quantum medium," have been employed in recent years to indicate the astounding energetic potential of what, for the past century, has been perceived incorrectly as nothingness.

Similar findings have been reached in biology. Molecular scientists have identified a striking phenomenon mentioned in the previous chapter—zero point energy—by which biological organisms use measurably more energy than is possible for them to receive from their daily intake of food, water, and air. This phenomenon occurs when the distance separating two non-charged surfaces, such as water and a cell membrane, becomes negligible, dimensional coherence takes place, lasing occurs and, by most indications, multidimensional torsion energy is drawn from the vacuum potential of space.

For those familiar with the officially acknowledged applications of Einstein's theories of Relativity, it should be easy to see how the mere existence of particles that move faster than light and/or communicate telepathically through an aetheric medium begins to unravel an entire paradigm for understanding the physical universe.[10]

On the subject of the "Crumbling of Certainty" in the wake of such logic-challenging discoveries in quantum physics, Charles Eisenstein in *The Ascent of Humanity* writes, "The whole idea of certainty of knowledge, built through objective reasoning, is only as sound as the objectivity at its basis. Question that, and we question the soundness of the entire edifice of experimentally-derived knowledge" on which our current sciences, and the worldview connected to them, depend.

Thus we find ourselves in the age of subquantum science, in which our greatest minds find themselves struggling to explain seemingly inexplicable phenomena having to do with apparently impossible, nonlocal events such as remote viewing and ESP—to cite two puzzling examples that now can be explained plausibly as intelligent movements of hyperdimensional torsion energy or universal creative consciousness. I propose that tachyons and related particle-waves are forms of

---

[10] Although it generally is assumed in Western scientific circles that Einstein's theories effectively do away with the need for a unified energy field, in the 1920s Einstein actually affirmed the contrary, stating that "in theoretical physics, we cannot get along without" such a field—which, interestingly, he referred to as "ether."

non-gravitationally bound or superluminal torsion energy emanating multidimensionally from our "transdimensional" point of origin at the center of our galaxy.

The Vedics created an elaborate science spanning yoga, meditation and diet for pooling this hyperdimensional energy, which they termed "prana," into their bodies. The Taoists developed a similar science for cultivating "chi." Early in the 20th century, Nikola Tesla theorized the existence of "scalar" waves (subsequently popularized by Tom Bearden) that transcend spatial limitations and are capable of acting instantaneously at a distance. Tesla created a prototype scalar system for free electricity using no generators or wires. Later, Wilhelm Reich became famous experimenting with "orgone" energy. Aether, *prana*, *chi*, scalar, orgone—all are names for the light-based aspect of the same spiritual or torsion energy that gave (and continues to give) rise to the holographic multiverse.

My research suggests that life as we know it depends on a double helix, structurally similar to DNA, of two differentiated, interfacing types of torsion energy: 1) hyperdimensional thought or *intention* manifesting as light; and 2) *sound* in higher-dimensional octaves which, like its counterpart, is a standing spiral wave capable of activating DNA, for example, with no time lapse across theoretically infinite distances.

To those wondering how such waves could contain sufficient information to (re)program DNA, it is worth remembering we live in a world crisscrossed with electromagnetic waves that carry highly complex television, telephone and radio signals that can be easily decoded with the right equipment. According to Richard Miller, a leading theoretician in the field of quantum bioholography, even more information can be encoded holographically than with simple electromagnetic encryption.

For practical purposes, the DNA molecule is designed brilliantly as a holographic torsion-wave-decoding biocomputer—one that magnetizes creative energy to it, and thus to our consciousness, that is aligned with our beliefs.

> **The DNA molecule is designed brilliantly as a holographic torsion-wave-decoding biocomputer—one that magnetizes creative energy to it, and thus to our consciousness, that is aligned with our beliefs.**

What if, by introducing healing sounds and intentions to the genome, sounds and intentions that derive from the same unified torsion energy of Source that I will argue in Part II is unconditional love, it is possible to key transposons in potential DNA to rearrange themselves and play the energy body in a higher octave, one more in tune with the unlimited creative consciousness of nonlocalized mind?

What if we thus can raise the harmonics of our system of electromagnetic fields and fully align ourselves with our Higher Selves? What if by raising our vibratory frequency from within, we can repattern our electromagnetic blueprint and seal the Fragmentary Body, allowing us to transcend limiting dualistic patterns and elements related to these patterns—physical, mental, emotional, and spiritual?

Again referencing the new Russian research in wave-genetics, Wilcock writes that "these studies give extremely convincing evidence that the DNA molecule is directly affected by outside energy sources. If DNA is actually assembled by an outside source of energy, then when we increase the flow of that energy into the DNA, we can also expect that the health and vitality of the organism will increase."

Basing his position on scientific studies demonstrating that the health of plants cannot be accounted for fully by their soil, air and water conditions, which suggests there is an overarching energetic component to their growth and development, Wilcock concludes that we "are left with the strong impression that torsion waves are the single most important factor in an organism's health."

Leigh and I think of ourselves not as "practitioners" or "healers," but merely as *facilitators* for the individual's own bioenergetic unfoldment. We consider Potentiation Electromagnetic Repatterning the first step in a four-part "rebirth cycle" that continues with Articulation Bioenergy Enhancement and Elucidation Triune Activation and culminates in Transcension Bioenergy Crystallization. This transformational process starts with a specific DNA activation that initiates a domino effect of electromagnetic repatterning that, like human gestation, takes just over nine months (42 weeks) to unfold.

Because there are two forms of hyperdimensional helical waves that interface with potential DNA, sound and light, Potentiation employs both. We use sound (produced vocally) and light (in the form of a non-directed healing intention, also language-based) to produce what we call an "energized narrative."

We contend that this conscious movement of torsion energy in the form of special words composed exclusively of vowels (see Part II) is capable of stimulating transposons in the human genome, activating a reset program that lies dormant in potential DNA. This program (which we do not need to create because it already exists) starts a chain reaction in which torsion energy works its way down

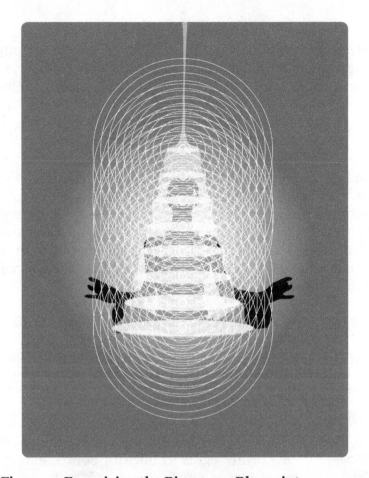

Figure 4: Energizing the Bioenergy Blueprint
The above image shows the flow of "torsion" energy or universal creative consciousness down through the electromagnetic fields that occurs after the Fragmentary Body is "sealed" at around the five-month mark following Potentiation. Utilizing the genetic sound-light translation mechanism, each sonic field, in turn, energizes the corresponding *chakra* with hyperdimensional light, which then transfers as bioenergy or *kundalini* to specific aspects of the subtle anatomy.

and up then down again through the various bioenergy centers on a gestational timeline.

After approximately five months, the electromagnetic fields and chakras recalibrate from nine to eight in number and the bioenergy vacuum constituted by the second field/chakra, the Fragmentary Body, seals itself. This occurs as the ninth and second fields/chakras fuse in what might be termed a sacred marriage of opposites.

Sealing, I cannot overemphasize, is an indispensable step on the path to true healing, as it lays the groundwork for a higher energy body—and ultimately bio-spiritual enlightenment—by initiating the process of integrating the fragmentation of the Self caused by duality (the Original Wound). In a profound sense, we can say that *sealing is required for healing or "wholing."*

Sealing is an indispensable step on the path to true healing, as it lays the groundwork for a higher energy body—and ultimately bio-spiritual enlightenment—by initiating the process of integrating the fragmentation of the Self caused by duality.

Over the next four months or so, this new bioenergy blueprint begins to fill with Source energy from the top down much like a tiered fountain as shown in Figure 4. The chakras slowly begin to open, becoming more powerful and efficient, as the electromagnetic fields gradually increase their harmonic resonance. At this point, "potentiators" often report a sense of integrating this new energy that can last up to an additional "gestation cycle" of nine months.

In terms of consciousness, Potentiation Electromagnetic Repatterning facilitates awareness of our true nature (which is divine) in relation to the Cosmos (which is holographic, meaning the part not

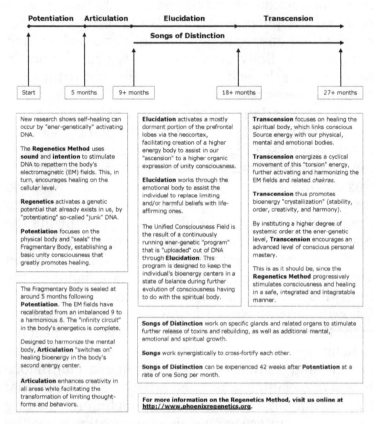

The Regenetics Method Timeline

| Potentiation | Articulation | Elucidation | Transcension |

Songs of Distinction

| Start | 5 months | 9+ months | 18+ months | 27+ months |

New research shows self-healing can occur by "ener-genetically" activating DNA.

The **Regenetics Method** uses **sound** and **intention** to stimulate DNA to repattern the body's electromagnetic (EM) fields. This, in turn, encourages healing on the cellular level.

Regenetics activates a genetic potential that already exists in us, by "potentiating" so-called "junk" DNA.

Potentiation focuses on the physical body and "seals" the Fragmentary Body, establishing a basic unity consciousness that greatly promotes healing.

The Fragmentary Body is sealed at around 5 months following **Potentiation**. The EM fields have recalibrated from an imbalanced 9 to a harmonious 8. The "infinity circuit" in the body's energetics is complete.

Designed to harmonize the mental body, **Articulation** "switches on" healing bioenergy in the body's second energy center.

Articulation enhances creativity in all areas while facilitating the transformation of limiting thought-forms and behaviors.

Elucidation activates a mostly dormant portion of the prefrontal lobes via the neocortex, facilitating creation of a higher energy body to assist in our "ascension" to a higher organic expression of unity consciousness.

Elucidation works through the emotional body to assist the individual to replace limiting and/or harmful beliefs with life-affirming ones.

The Unified Consciousness Field is the result of a continuously running ener-genetic "program" that is "uploaded" out of DNA through **Elucidation**. This program is designed to keep the individual's bioenergy centers in a state of balance during further evolution of consciousness having to do with the spiritual body.

Transcension focuses on healing the spiritual body, which links conscious Source energy with our physical, mental and emotional bodies.

Transcension energizes a cyclical movement of this "torsion" energy, further activating and harmonizing the EM fields and related *chakras*.

Transcension thus promotes bioenergy "crystallization" (stability, order, creativity, and harmony).

By instituting a higher degree of systemic order at the ener-genetic level, **Transcension** encourages an advanced level of conscious personal mastery.

This is as it should be, since the **Regenetics Method** progressively stimulates consciousness and healing in a safe, integrated and integratable manner.

Songs of Distinction work on specific glands and related organs to stimulate further release of toxins and rebuilding, as well as additional mental, emotional and spiritual growth.

Songs work synergistically to cross-fortify each other.

Songs of Distinction can be experienced 42 weeks after **Potentiation** at a rate of one Song per month.

For more information on the Regenetics Method, visit us online at http://www.phoenixregenetics.org.

Figure 5: The Regenetics Method Timeline
The above chart delineates the Timeline for the four primary DNA activations of the Regenetics Method.

only reflects, but also contains the whole), empowering individual discernment to begin attracting higher "quantum outcomes" following the universal law that "consciousness creates."[11]

This productive use of awareness invites one to begin facing limiting thoughts, emotions and beliefs and by itself can lead to markedly increased vitality, more balanced interactions with one's

---

[11] See Part II.

environment (i.e., fewer allergic reactions), greater financial abundance, more loving relationships, even renewed life purpose. Many also have reported that raising their ener-genetic resonance and sealing their Fragmentary Body appear to be critical factors in releasing deeply held cellular toxicity and pathogens, including parasites.

Articulation Bioenergy Enhancement is appropriate as of the five-month mark following Potentiation (see Figure 5), after the electromagnetic fields have recalibrated and sealing has occurred. At this point, a new bioenergy blueprint (with no interruption or leakage in the form of the Fragmentary Body) is in place to begin utilizing the potent life force Articulation gently stimulates.

In Vedic teachings, this life force is called *kundalini*. Leigh and I understand kundalini as the individual's own torsion life-wave that, again following "As above, so below," marries the macro- with the microcosmic in the process of healing. Articulation safely "switches on" kundalini at the genetic and cellular levels, providing a continuous bioenergy supply for creativity and personal transformation in all areas—including artistic expression, interpersonal communication, healthy sexuality, and rebuilding through diet and exercise.

Elucidation Triune Activation, the third phase of the Regenetics Method, is appropriate following Articulation as of the nine-month (42-week) mark of Potentiation (Figure 5). Elucidation appears to stimulate a mostly dormant portion of the prefrontal lobes by way of the neocortex and triune brain, facilitating creation of a higher energy body.

Elucidation also encourages transformation by assisting the individual to replace limiting and/or harmful beliefs with life-affirming ones. This restructuring of the belief system can change one's experience of reality quite dramatically. The movement here is toward embodiment of a new awareness based no longer on duality and separation, but on unity consciousness and unconditional love.

The fourth and final phase of the Regenetics Method, appropriate as of the 42-week mark following Elucidation (Figure 5), Transcension Bioenergy Crystallization focuses on healing distortions in the spiritual body (the most fundamental and powerful level of our subtle anatomy) in order to facilitate an advanced level of conscious personal mastery—briefly defined as unconditional love of oneself as simultaneously the Creator and the created extended outward to all perceived others.

Developed for the true spiritual seeker or "adept," Transcension stimulates greater movement of kundalini as the bioenergy centers are brought to an even higher level of organization or "crystallization," facilitating luminous embodiment of divine consciousness. This ultimate level of healing acknowledges that there is no difference between self and other since, in the grand scheme of creation, there is only the One. The fundamental, reality-altering idea here is that it is only by knowing the authentic Self, and then allowing that Self to express its wholeness, that we heal the world.

8
A Practical Application of Era III Medicine

By way of closing this section, I wish to return briefly to the work of Larry Dossey and address one practical application of Era III medicine. Dr. Dossey writes, "Recently modern scientists have discovered that nonlocal events, meaning events that don't happen where they're initiated necessarily, are not fantasy but are part of the fabric of the universe." Citing well-known physicist Nick Herbert, Dossey enumerates three primary characteristics of nonlocal events. Nonlocal events are:

1) *unmediated*, meaning you do not necessarily need a transmission field for a nonlocal event to occur, that it occurs even in the absence of an identifiable field;

2) *unmitigated*, meaning their strength does not dwindle because of distance (or time); and

3) *immediate* (and sometimes even anachronistic, meaning they can happen before they happen).

On many occasions, Leigh and I have found that our clients have begun to feel Potentiation Electromagnetic Repatterning happening *before* the actual session. Time and space seem irrelevant with the Regenetics Method, which is what one would expect working with torsion energy capable of moving faster than observable light. In fact, a principal characteristic of the healing applications of wave-genetics, as explained by Gariaev, is quantum nonlocality.

Here again, Gariaev's research provides a plausible explanation. The Russian team found that DNA can cause "disturbance patterns" in space, generating small electromagnetic wormholes of a subquantum nature. These microscopic, DNA-activated wormholes, similar in their ability to bend light to the nonlocal energy signatures found in the vicinity of black holes, including Galactic Center, are connections between different areas in the multiverse through which data (such as sound and light codes designed to assist human evolution) can be transmitted outside the space-time continuum. *Distance healing, for example, thus is reasonably accounted for as a meta-genetic phenomenon.*

Finally, after much research Leigh and I formulated our ideas and performed the first Potentiation on ourselves while still living in South America. Initially, to be perfectly honest, I did not experience any significant shifts. But after a few weeks, I realized I was beginning to crave foods I had been unable to eat for years. Starches mostly, carbohydrates.

Leigh theorized I was craving starches again because my body was starting to cleanse and was asking for foods to help bind toxicity from my cells and escort it out of my system without further damaging my tissues. She suggested that I try some of the foods I had been craving such as pasta, bread, rice, and potatoes. So I began eating them, timidly at first but with increasing gusto as it became obvious they no longer were overrunning me with Candida or causing unbearable bloating.

If Potentiation had stopped there, we both knew we had been given an extraordinary gift. Nothing else had been able to restore my ability to enjoy the full range of foods.

But Potentiation did not stop there. I began getting back greater and greater levels of vitality and was able to start exercising again. Within six months, I returned to swimming a mile without stopping. I had been unable to perform anywhere close to this for nearly seven years. Years later, while still being able to eat and drink whatever I please, I can swim and exercise with remarkable stamina for my age.

I began getting back greater and greater levels of vitality and was able to start exercising again. Within six months, I returned to swimming a mile without stopping. I had been unable to perform anywhere close to this for nearly seven years.

My body was recovering its inherent wisdom, that had been undermined by vaccines, and was beginning to harmonize with the natural environment again. The experience reminded me of a story from Greek mythology about a giant named Antaeus who wrestles Hercules.

Antaeus derived his great strength from his connection with Earth. Every time Hercules tried to pin him, Antaeus would simply touch the ground and grow exponentially stronger. Finally, when he was almost spent, Hercules realized the only way to defeat Antaeus was to lift him off the ground where he could no longer draw strength from Earth. This was how Hercules strangled and killed Antaeus.

A month after Potentiation, for the first time in nearly eight years, I felt as if Hercules finally had set me back on the ground. I knew intuitively, as well as experientially, that I again was drawing strength from food, water, and sunlight. Something altogether profound was happening in me. Leigh also underwent an ener-genetic transformation with many tangible results that, to our amazement, included the partial straightening of her lifelong scoliosis that not even a back brace, regular chiropractic treatments and intensive Rolfing® had been able to modify.

We wanted to share Potentiation with family and friends, but were still living on another continent. One of my mentors suggested we could do this work at a distance using the elements we already were using: sound and intention. This struck me as very much in line with Dossey's notions of Era III medicine and nonlocalized mind, to say nothing of the theories and applications of wave-genetics.

We began to explore the possibility of performing Potentiation at a distance. One technique we looked to for ideas was radionics. Radionics is a type of energy medicine that historically has been performed remotely.

Radionic photography is a concrete example that thought (intention) is a form or function of torsion waves manifesting as light.

There is some very substantial proof of the radionic connection between practitioner and patient in something called "radionic photography." This is where film that never actually is put in a camera, being exposed only to the thoughts of the radionics practitioner, is then developed without applying light to the film. Radionic photography is a concrete example that thought (intention) is a form or function of torsion waves manifesting as light.

The matriarch of radionics, a chiropractor named Ruth Drown, created photographs of people at a distance by simply connecting with them mentally. She produced some uncanny images. In one she captured a fetus inside the womb of a patient who lived miles away. The shape of the fetus is clearly recognizable and the suggested anatomy of the mother seemingly correct.

Drown made other radionic photographs "taken" sometimes hundreds of miles away, including one distinctly showing surgical tools entering recognizable organs. These two amazing images have been published in Tansley's *Radionics: Interface with the Ether Fields.*

Gariaev's research in the various healing applications of wave-genetics, which have been documented to work remotely, strongly supports the idea that DNA constitutes a "network" comparable to the Internet that, being present anywhere, is simultaneously present everywhere—effectively doing away with distance.

Also a radionics practitioner, Tansley explains such remote energy transmission in terms of the "psi-field," described as a matrix in space filled with triangular energy vortices that allow for transmission and reception of intention or thought (torsion) energy between distant places. This takes us back to spiral standing waves that transcend time and space and is very much in keeping with the holographic model.

Alternatively, Gariaev's research in the various healing applications of wave-genetics, which have been documented to work successfully at distances of up to twenty kilometers, strongly supports the idea that DNA constitutes a "network" comparable to the Internet that, being present anywhere, is simultaneously present everywhere—effectively doing away with distance.

Biologist Rupert Sheldrake's Morphic Resonance theory posits a similar transpersonal "morphogenetic" network. Dr. Sheldrake's concept of "formative causation" emphasizes the existence of "morphic fields" that unite entire species universally outside space-time. According to Sheldrake, these omnipresent resonance fields can be expressed biologically if correctly tuned into—for example, through DNA—even if a species is extinct.

Some readers also may be familiar with the notion of "noosphere," the name given by the great Jesuit philosopher Pierre Teilhard de Chardin to the field of mind and thought that encircles the planet and enables zeitgeist to happen: spontaneous transfer of ideas and technologies that suddenly seem to leap from consciousness to consciousness. Lynne McTaggart's popular book *The Field: The Quest for the Secret Force in the Universe* provides yet another take on this nonlocal unifying fabric.

The final piece in the development of Potentiation Electromagnetic Repatterning fell into place when we were led to study the healing techniques of indigenous Hawaiian shamans known as *Kahunas*, who can heal people miraculously, often at a distance, using intoning. We learned that Kahunas, in order to transmit healing sounds to an ill person by proxy, use Earth's ley lines, which they call "aka threads"—a comprehensive network of sensitive "fiber-optic" graphite veins that seem almost designed to communicate between carbon-based life forms across distances.

In essence, Kahunas transmit sound through the earth while "triangulating" on the ill person with the light of their intention. It might even be said they generate an *energized narrative* sent to the recipient by sound and intention. This balanced use of telluric and astral energies almost certainly activates a self-healing potential in the individual's DNA, even though distance intervenes.

Although Leigh and I have come to recognize that Regenetics activations "travel" via the "information superhighway" through which DNA is connected to (and can be modified by) nonlocalized torsion energy, these concepts involving psi-fields and aka threads were very timely, informative, and inspirational.

Before returning from South America, we had the opportunity to test our theories about distance DNA activation and electromagnetic repatterning with approximately thirty people living in the United States. From that group, we compiled our first list of fourteen Testimonials that went for several pages. These altogether striking Testimonials included one spontaneous remission of a chronic rash; several cases of food allergies dramatically improving or disappearing; a shared sense of greater vitality; two reports of insomnia going away; and significant positive shifts in depression and fatigue.

Since then, while continuing to offer Potentiation to thousands of clients worldwide, we have integrated three additional DNA activations— Articulation Bioenergy Enhancement, Elucidation Triune Activation, and Transcension Bioenergy Crystallization—that round out the Regenetics Method. The results have been phenomenal, often surpassing our wildest expectations. This is especially the case when we receive enthusiastic feedback from clients living in different parts of the globe whom we have never met, as we regularly do.

You probably can imagine that Regenetics could be a challenging concept for many people today, that it might initially strike the more empirically minded as absurd. But like any revolutionary truth whose time has come, we believe that in the very near future the Regenetics Method, and similarly profound modalities based on emerging scientific principles supported by indisputable evidence, will be acknowledged gradually, then accepted universally.

• PART II •
SACRED COSMOLOGY,
SACRED BIOLOGY:
THE REGENETICS METHOD & THE
EVOLUTION OF CONSCIOUSNESS

9
The Shift in Human Consciousness

Are you aware that a Shift in human consciousness is occurring even as you read these words that employs celestial triggers such as supernovas and Earth's alignment with Galactic Center in the years leading up to 2013 to trigger the evolution of our species? This is perhaps why, consciously or otherwise, the Regenetics Method has piqued your interest.

This Shift has been documented in a stimulating multimedia presentation entitled "Preparing for the Shift," by Barry and Janae Weinhold, Ph.D.s. Over decades devoted to the study of consciousness and evolution, the Weinholds, both trained psychologists, have gathered overwhelming evidence that humanity is in the middle of a long-awaited Shift in consciousness predicted in hundreds of indigenous cultures all over the world.

Today this Shift is visible nearly every time you open a newspaper, turn on the TV, or check your email. It can be seen in the breakdown of many old structures such as those that underpin governments, banks, corporations and churches, as well as in families, couples, and individuals. It is also evident in the ecological breakdown of numerous Earth systems, a widespread perception time is accelerating, drastic changes in weather patterns (global warming), more people feeling overwhelmed by modern life's complexity, and increased polarization between groups, religions, and regions.

Fortunately, along with signs of breakdown, the Weinholds emphasize there is also considerable evidence of *breakthrough*: the appearance of stunningly gifted children in unprecedented numbers, the emergence of innovative and integrated healing modalities, people becoming less "religious" and more "spiritual," and the dawning of new communities and social structures based on servant leadership and other "win-win" partnership principles.

The significance of the winter solstice on December 21, 2012, according to the Mayan, Aztec, Incan and Hopi traditions, is that this date marks the close of several cycles of time.[12] The first is the end of the 26,000-year (25,625-year) Mayan calendar, also called the "precession of the equinoxes" and the *Annus Magnus* ("Great Year"), considered by many a gestation or birth cycle for Earth. The Great Year charts our lengthy, cyclical journey through the twelve constellations of the zodiac.

Interestingly, the number 26,000 is very close to Plato's "ideal" number of 25,920, as well as to the length of one third-dimensional evolutionary cycle (25,000 years) as described in the unparalleled body of intuitive teachings known as *The Law of One*. Mayan timekeepers believe that human evolution unfolds as a result of such precisely calibrated master cycles of time. They predict that Earth and humanity are about to be birthed into a new reality based on

---

[12] There is some spirited disagreement among scholars as to the exact date of the end of the Mayan calendar. In *Maya Cosmogenesis 2012*, John Major Jenkins demonstrated that on December 21, 2012, Earth's "precessional" axis will be aligned directly with Galactic Center. Carl Johan Calleman proposes the alternative date of October 28, 2011, which theoretically could be more accurate based on evolutionary cycles as opposed to astronomical data. See *The Mayan Calendar and the Transformation of Consciousness*.

unity predicated on a dramatic advance in consciousness.

From a Mayan perspective, the Weinholds ask, "What began roughly 26,000 years ago?" Their extensive psychohistorical research indicates this marked the beginning of humanity's psychological *individuation*.

Mayan timekeepers believe that human evolution unfolds as a result of precisely calibrated master cycles of time. They predict that Earth and humanity are about to be birthed into a new reality based on unity predicated on a dramatic advance in consciousness.

In human terms, becoming "individuated" means moving from being unconsciously united with the Creator or Ground of Being; to choosing to become divided from the Creator and developing separate individual consciousness; to finally returning to the Creator as conscious, aware individuals. Once people fully individuate, it becomes possible for them to make empowered, discerning choices and use intention to co-create reality with Source—or perhaps more accurately, to create reality *as* Source.

The second cycle of time ending around 2012 highlighted in "Preparing for the Shift" is the close of the Galactic Year. It takes roughly 225 million Earth years for the Milky Way Galaxy to make one complete rotation in the sky, which is believed to be a birth cycle for our galaxy (Figure 6). From a galactic perspective, the Weinholds ask, "What was conceived on Earth 225 million years ago?"

They point out this was when Earth's landmass, known as Pangaea, began separating into what we now know as the seven continents. This process of planetary individuation not only correlates with continental drift theory; as mentioned in Part I, there is also an energetic correspondence between Earth's twelve tectonic plates responsible for continental drift and the twelve pairs of cranial nerves in the human brain, which are linked to the biblical Twelve Tribes.

Considered together, this evidence suggests that Earth, like humans, has undergone her own "separating out" as a precondition for individuation. Based on such interconnectedness, it is also reasonable to expect that as human consciousness exponentially increases leading up to the 2011-13 "window," Earth also will undergo a significant—and observable—transformation.

Astronomers studying Galactic Center report that it periodically becomes extremely active. During these episodes, it releases fierce barrages of cosmic energy equal to thousands of supernova detonations. These outbursts are the most supercharged phenomenon in the known universe.

A growing number of researchers such as Sergey Smelyakov, author of a fascinating paper entitled "The Auric Time Scale and the Mayan Factor," in addition to many indigenous peoples worldwide and an increasing number of intuitively derived sources including *The Law of One*, theorize that as Galactic Center becomes more energized, it catalyzes human evolution through frequency emissions of torsion waves transmitted to Earth via the sun.

A comprehensive scientific model for the "Energetic Engine of Evolution" has been proposed by David Wilcock, a gifted speculative scientist whose theory of "Evolution as 'Intelligent Design'" deserves summarizing here.

Citing the work of a vanguard of researchers including Tim Harwood, Glen Rein, Bruce Lipton, Richard Pasichnyk, Aleskey Dmitriev, Vladimir Poponin and Peter Gariaev, Wilcock presents a model that unites many disciplines and provides several critical missing pieces to the evolutionary puzzle. In his own words, this provocative model "suggests that humanity is on the verge of a near-spontaneous metamorphosis into a more highly evolved state of consciousness."

Basing his analysis on the realization, embraced by more and more of today's scientists, that Darwinian evolutionary theory is "extinct," Wilcock observes that the "probabilities that DNA could evolve by 'random mutation' are so minute as to be utterly laughable—akin to the idea that if you have enough monkeys tapping away on typewriters, one of them will eventually produce a complete Shakespearean play."

Far surpassing the reach of gradual, incremental evolution, which certainly occurs as environmental *adaptation*, the fossil record from all over the planet makes it abundantly clear that species regularly *evolve* in heretofore inexplicable leaps and bounds, skipping what would seem from a Darwinian perspective to be crucial evolutionary phases. At the top of a long list of species whose evolution has baffled science is the human species.

Although for more than a century a "missing link" has been assumed to exist based on largely unchallenged Darwinian presumptions, scientists have never managed to discover it. "When we consider that the size of the brain literally doubled between that of humanity's apparent ancestors and ourselves, with no evidence of a smooth transformation whatsoever," writes Wilcock, "once again we see a spontaneous evolution of the creatures on Earth."

One scientist associated with *National Geographic*, studying the intricate bone carvings dating to 70,000 B.C. found at Blombos Cave in South Africa, concluded that behavioral evolution mirrors anatomical development—an important observation meaning, in Wilcock's words, that "spontaneous evolution is not simply physiological, but consciousness-related as well. When a new bodily form has emerged, consciousness changes appear to occur."

Physicist Amit Goswami, author of *The Self-aware Universe: How Consciousness Creates the Material World*, advances a parallel theory of evolution in which consciousness is primary and matter is secondary. Goswami similarly points out that natural selection simply cannot account for significant changes in the complexity of systems—as in, from primate to human. "Instead," he writes, such changes "show the quantum leap of a creative consciousness choosing among many simultaneous potential variations."

Another forward-thinking scientist on a similar wavelength is physicist Lee Smolin. In *The Life of the Cosmos*, Smolin writes that the "idea that the laws of nature are immutable and absolute ... might be as much the result of contingent and historical circumstances as they are reflections of some eternal, transcendent logic."

His intriguing alternative to this fixed viewpoint is "cosmological natural selection," where black holes grant access to different areas in the multiverse where cosmological conditions are at least somewhat different. The student of *The Law of One* should have no trouble seeing how this is an uncannily accurate description of the creation of new "octaves" of existence after the evolutionary potentials of previous octaves have been explored and integrated.

Moreover, as indicated by the Mayan calendar and very much supported by the cosmology elaborated in *The Law of One*, rather than in fits and starts, evolutionary fast-forwards of consciousness and physiology happen in organized, predictable cycles. Theorizing a "harmonic relationship" between the 26,000-year Mayan calendar and 26-million-year period between extinctions/evolutionary leaps in the fossil record, Wilcock notes that all Earth species suddenly have evolved, or metamorphosed, every 26 million years, making a strong case for "an outside energetic influence that operates in a regular, cyclic fashion."

Here as usual, Wilcock's seemingly radical theory is founded impeccably in research—this time, that of University of Chicago paleontologists David Raup and John Sepkoski, who discovered the 26-million-year cycle in the fossil record during the most complete study of marine fossils ever performed that examined over 36,000 different genera. Raup and Sepkoski's research, which reveals regular cycles of "punctuated equilibrium" involving spectacular examples of speciation, clearly shows unprecedented creatures spontaneously appearing in the layers of the ocean's crust over time.

"Most people simply write this off as asteroid/comet collisions or volcanic eruptions" causing these evolutionary jumps, writes Wilcock in his blog. "However, some of [these examples of speciation] occurred without any cataclysms going on. Living things, all over the earth, simply 'decided' it was time for a change. Or ... something pushed them!"

More recently, Robert Rohde employed this same set of data to uncover another cycle of much greater scope: 62 million years. While Raup and Sepkoski's cycle only extended halfway back to the start of life on Earth (about 250 million years), Dr.

Rohde's cycle reaches all the way back to the appearance of life on our planet—approximately 542 million years ago. "Before skeptics hurl their inevitable brickbats," comments Wilcock, "let's be advised that Dr. Rohde's study was published in perhaps the single most prestigious, top-drawer science publication in the world ... *Nature.*"

To answer the question what outside energetic influence is responsible for rhythmic evolutionary revolutions in species, it is necessary to factor in the concept of torsion energy introduced in Part I.

Some writers, most notably Barbara Hand Clow, have focused attention on something called the Photon Belt or Photon Band, which can be envisioned as a torsion-wave "light lattice" connecting Earth via our sun to Galactic Center that serves as a guiding data communication network for human and planetary evolution. Such an understanding of the universal communication network is deeply "shamanic," underpinning the majority of the world's native wisdom traditions, as well as the Regenetics Method.

While some astronomers have scoffed at the notion of a Photon Band, other scientists who grasp the higher-dimensional nature of this network's light understand that it not only exists, but plays a critical role in cosmic evolution. In 1962, writes Ron Radhoff in *New Science News*, the year many

> say we entered the Aquarian Age, we began to enter into the influence of [the] photon-belt ... We will pass into the center of it by the year 2011 ... St. Germain refers to the photon-belt as the Golden Nebula, a parallel universe of much higher vibration. Little by little it is absorbing our universe. As we merge with this higher vibration universe, it will become the catalyst for massive changes.

Figure 6: The Photon Band & Black Road
It takes roughly 225 million Earth years for the Milky Way Galaxy to make one complete rotation through the Photon Band, which some believe to be a birth cycle for our galaxy. The Black Road can be conceptualized as simultaneously an astrophysical and "meta-genetic" alignment with Galactic Center that engenders a Shift in consciousness and physiology, allowing humanity to return "home."

In Ken Carey's visionary writings, the Photon Band is said to be composed of various "creation beams," a phrase that anticipates both Mayan calendar scholar José Argüelles' notion of the "Galactic Synchronization Beam" and scientist Paul LaViolette's "Galactic Superwave" (minus the paranoid interpretation of this natural phenomenon infusing the latter). "In the course of its movement through space each galaxy passes through successive *creation beams* ... designed to encourage processes relating to certain desired patterns of structural development," writes Carey. "Each of these regions establishes, through the quality and nature of its frequencies, a vibrational climate designed to elicit a specific type of creation."

"As this arm of the Milky Way galaxy has rotated slowly through this past creation beam," continues Carey, "the vibratory climate of that beam has encouraged humankind's emergence, multiplication, and flourishing ... But [our] planet's continued physical movement through space is now bringing it out of that creation beam and out of the vibratory conditions [we] have known historically" into a new level of vibration designed to promote the emergence of more evolved life-forms.

Such an understanding of varying levels of vibration in different areas of the galaxy, and their intimate relationship with spontaneous speciation, as evidenced in the fossil record, is a hallmark of *The Law of One*, where what I am calling the Photon Band is described meticulously as operating in virtually clocklike fashion to promote spectacular jumps in conscious evolution. A central concept in *The Law of One* is that such evolutionary leaps are indeed leaps, being "graduations" from one "grade level" to the next which typically occur within a single generation.

Of humans it is stated, "Those who, finishing a cycle of experience, demonstrate ... understanding ... will be separated by their own choice into" an appropriate and comfortable dimension of experience. "All are harvested regardless of their progress, for [by this time] the planet itself has moved through the useful part of that dimension and begins to cease being useful for ... lower levels of vibration."

Wilcock's encyclopedic research on this heretofore fringe subject takes into account reams of NASA and other scientific data proving that unprecedented climate changes (global warming) are occurring not just on Earth, but throughout our solar system. Based on this hard data, as well as extraordinarily consistent evidence derived from a plethora of intuitive sources, *the phenomenon of "heating up" observed literally all around us is a direct result of our entry into a more energized area of space.* Logically, this goes a long way toward proving that something like a Photon Band exists as regular patterns of hyperdimensional torsion radiation emanating from Galactic Center.

Both Wilcock and Clow envision the Photon Band as tracing figure-eights throughout the spiraling layout of the Milky Way Galaxy (Figure 6). It appears from Dr. Nikolai Kozyrev's research involving aether mentioned in Part I that such looping energy based on the Phi ratio (1.6180339...) is, among other things, directly responsible for our cyclical experience of time. On this subject, the title of Gregg Braden's latest book speaks volumes: *Fractal Time: The Secret of 2012 and a New World Age.*

This same fractal, spiraling torsion energy also directly creates our DNA and imprints its inherent Phi ratio on practically all of nature's systems, from the way our hair swirls atop our heads, to the growth patterns of leaves on plants, to the unmistakable shape of many seashells. As Geoff Ward observes in *Spirals: The Pattern of Existence,*

The spiral is the sign of the eternal, creative, unifying and organizing force or principle at work in the universe, and especially of the ongoing creation of consciousness. It is a divine mark on nature, what may be termed God's personal signature on the cosmos, the Great Architect's own autograph—from the cradles of stars and planets in the awesome spiral arms of galaxies to the beautiful double helix structure of the DNA molecule [...] As the spiral seems to be integral to physical growth, so it is also the symbolic pattern of human spiritual growth ... [T]he spiral is as much part of our "cultural" DNA as it is part of our biological DNA [...] Indeed, I have come to realize that "spirality"—the condition of being spiral—and "reality" are almost interchangeable terms.

Recalling the aether theories of Kozyrev, fellow Russian scientist Sergey Smelyakov's research demonstrates that the harmonic vibrations of Phi, also referred to as the Golden Mean and Fibonacci sequence, inform the very fabric of space-time. Mathematically, the Photon Band appears to be structured on Phi, producing set cosmic intervals the Maya apparently were aware of when constructing their uncannily accurate calendar. Smelyakov's "The Auric Time Scale and the Mayan Factor" compellingly suggests that Earth connects to Galactic Center via our solar system in a harmonic, fractal fashion he calls "Solar-planetary Synchronism," a vibratory relationship based on the Golden Mean.

In an article entitled "The Ultimate Secret of the Mayan Calendar" reprinted in *DNA Monthly*, Wilcock analyzes Smelyakov's research, writing that it helps explain the end of the Mayan calendar in geometric terms as an "infinitely-converging end point" in which time appears to "collapse." This is because time as we experience it follows the imploding spirals of Phi much like a finger tracing the cyclical involutions of a conch shell to its center point.

A salient point in the evolutionary model I am elaborating here is that evolution of species ultimately is driven not by material, but by metaphysical energy, or *consciousness*, of a spiraling, meta-genetic nature.

> ***Evolution of species ultimately is driven not by material, but by metaphysical energy, or consciousness, of a spiraling, meta-genetic nature.***

Swedish Biologist Carl Johan Calleman reaches precisely the same conclusion in *The Mayan Calendar and the Transformation of Consciousness*, cautioning "against a narrow interpretation of the term *cyclical*. The Mayan calendar describes evolutionary rather than strictly cyclical processes, so history is more like a spiral of evolution, in which similar types of events are favored at certain points in the cycle."

Calleman is quick to clarify that the "results of these cyclical bursts of creativity are never identical; a repetition of identical cycles does not generate evolution. History is rather a process resulting from stepwise increasing levels of consciousness."

Reading between the lines, the experienced student of the Mayan calendar will hear in Calleman's words a critique of interpretations of the Mayan calendar as purely a measurement of revolving astronomical cycles—the type of "materialistic" reading given to the Mayan calendar by Braden and John Major Jenkins, for example. Instead, Calleman's Mayan calendar is, first and foremost, a map of the *evolution of consciousness*. Driving home his point, Calleman emphasizes that "endlessly repeated identical astronomical cycles could never explain the evolution of consciousness."

On the subject of utilizing the Mayan calendar as a spiritual map with a clearly defined teleological focus, Calleman writes, "If today we are to embrace a worldview in which consciousness is more important than matter, we ... need to base our timekeeping on the nonphysical, invisible reality [that gives rise to reality] rather than on the physical."

History, then, does not exactly repeat itself; it is more like climbing a spiral staircase—one with pronounced evolutionary changes in consciousness (and body types) as we ascend from floor to floor.

Terrence McKenna's well-known "Timewave Zero" theory has many similarities to Calleman's reading of the Mayan calendar as a progression through various "Underworlds" of consciousness, as well as to Smelyakov's Solar-planetary Synchronism. In all three of these models, which to the best of my knowledge were developed independently of one another, evolution of consciousness occurs in exponentially increasing phases of intensity and velocity as we approach 2012, culminating in a virtually instantaneous transformation of reality.

Torsion energy spiraling as the Photon Band from the creative consciousness at the Core of our universe is Wilcock's "Energetic Engine of Evolution." Because of its curvilinear form, the Photon Band is composed of swaths of lesser and greater density of torsion waves manifesting as hyperdimensional light. As our solar system orbits episodically into galactic regions characterized by greater density of torsion waves (i.e., greater light or consciousness), which it is doing currently, life on our planet, including the living organism that is Earth, is stimulated intelligently to evolve in spectacular ways not only physically, but also mentally, emotionally, and spiritually.

"By combining the effects of geo-cosmic change with the overall flourishing of humanity in the cultural and spiritual sense," observes Wilcock, "we see that as the cycle continues to exponentially accelerate its energetic rate of vibration into the 2012-2013 'singularity,' we can expect ... rapid increases in human awareness." This centripetal cycle leads inexorably to a "discontinuous mega-event where 'time and space collapse.'" Perhaps this transformation of our experience of time and space is the truth behind the disjointed description of the "end of days" in the Book of Revelation.

As our solar system orbits into galactic regions characterized by greater density of torsion waves (i.e., greater light or consciousness), which it is doing currently, life on our planet, including the living organism that is Earth, is stimulated intelligently to evolve in spectacular ways not only physically, but also mentally, emotionally, and spiritually.

In humans, evolutionary activation occurs as torsion waves stimulate potential DNA's transposons to rewrite the genetic code—a phenomenon supported by a considerable amount of scientific data. Bruce Lipton's research in epigenetics, as detailed in *The Biology of Belief: Unleashing the Power of Consciousness, Matter, and Miracles*, unambiguously affirms that cells possess the ability to reprogram their own DNA, with measurable physical results such as otherwise inexplicable dietary modifications in organisms, when prompted environmentally. Dr. Lipton hypothesizes that such rewriting, which is typically beneficial, accounts for up to ninety-eight percent of evolutionary transformation.

Similarly, in a concise but excellent study entitled "Retrotransposons as Engines of Human Bodily Transformation," biochemist Colm Kelleher addresses the subject of radical genetic evolution as a result of what he terms a "transposition burst."

"If one were to hypothesize a transmutation of the human body," writes Dr. Kelleher, it

> would be necessary to orchestrate a change, cell by cell, involving the simultaneous silencing of hundreds of genes and the activation of a different set of hundreds more. A transposition burst is a plausible mechanism at the DNA/RNA level that could accomplish such a genome wide change. Transposition bursts comprise the concerted movement of multiple mobile DNA elements from different genetic locations to new positions, sometimes on different chromosomes ... Human DNA contains an abundance of the necessary genetic structures to accomplish a transposition burst involving hundreds, or even thousands, of genes.

Referencing a particular DNA sequence containing three different transposon families (a genetic trinity) arranged in beadlike formation, Kelleher theorizes that owing to its tripartite configuration, this DNA sequence would be "an effective participant in large scale transposon mediated genetic change that eventually results in transformation of the human body."

Perhaps the most undeniable evidence supporting the concept of a torsion life-wave or Photon Band of universal creative consciousness "meta-genetically" directing the spontaneous formation and development of Earth species comes from Tim Harwood, who calls attention to one of nature's more miraculous phenomena.

After caterpillars form their chrysalis during metamorphosis, it is a little-known but very relevant fact that they completely dissolve into a soup of amino acids before reassembling into butterflies. This soup contains no recognizable cells or DNA as it is understood presently, but when the time is right, the torsion life-wave signals the DNA to recombine and, within a matter of days, cells emerge to create new life-forms (Free).

Wilcock concludes that the human species, somewhat like caterpillars entering metamorphosis, currently is "being programmed by the galactic center to become more advanced while ... still here in our bodies." This is made possible because the

> DNA molecule is like a programmable piece of hardware ... so that if you change the energy wave that moves through it, the jumping DNA will encode it into a completely different form. It is therefore possible that as we move into increasingly "intelligent" zones of energy in the galaxy, the DNA energy patterns for the creatures on the planet are all upgraded, and the mutations occur so rapidly—well within one lifetime—that no "transitional" fossils exist.

Earth's movement through a denser area of the Photon Band aligned with Galactic Center may have begun as far back as the 1930s but certainly by 1980; increased greatly in intensity around the time of the so-called Harmonic Convergence in 1987; will enter into a historic astronomical alignment around 2012; and will be complete (from our present linear perspective) by about 2016.

Over the course of the past two decades, as Wilcock and the Weinholds point out, major Earth, planetary and solar changes—from unprecedented alterations in planetary atmospheres to drastic surges in volcanic and earthquake activity—have been observed. Arguably of greatest significance from our perspective is that the sun now is moving into direct alignment with Galactic Center. During this transit, the sun's magnetic field has increased over 230% and there have been wild swings in levels of sunspot activity (as reported by NASA and other space agencies) that at times have transmitted record-breaking waves of electromagnetic (to say nothing of torsion) energy to Earth and, thus, to us.

It is worth noting we are composed of the same substances found in the heavens, so really it is not so odd that celestial events should impact us profoundly. Harvard professor of astronomy Robert Kirshner has remarked that "supernovas created the elements we take for granted—the oxygen we breathe, the calcium in our bones, and the iron in our blood are products of the stars."[13] Other researchers, observing that most of DNA's amino acids also are found in space, have hypothesized that DNA actually came from space—an increasingly popular theory known as "Panspermia."

Fascinatingly, Fritz Albert Popp's research in biophotons describes the dying process of cells as virtually identical to that of stars. Just before dying, cells transform into "supernovas" as the light they emit increases in intensity a thousand times before suddenly being extinguished.

In 2005 NASA's Spitzer Space Telescope identified some of life's key ingredients in the dust spiraling around a developing star. These gaseous precursors to DNA's building blocks were discovered in the star's terrestrial planet zone, where planets such as ours are believed to form. The findings mark

---

[13] Quoted in William Henry, *The Healing Sun Code*.

the first time these substances (hydrogen cyanide and acetylene) have been found in another terrestrial planet zone besides ours.

"This infant system might look a lot like ours did billions of years ago, before life arose on Earth," commented Fred Lahuis of the Netherlands' Leiden Observatory and the Dutch space research institute SRON. Lahuis and fellow researchers were surprised to detect the organic (carbon-based) gases around a star known as IRS 46 in the Ophiuchus or "Serpent Bearer" constellation approximately 375 light-years away.

Galactic Center, our point of origin, also is located in Ophiuchus. Writing on the subject of 2012 in *Galactic Alignment: The Transformation of Consciousness according to Mayan, Egyptian, and Vedic Traditions*, John Major Jenkins explains that Ophiuchus was a key player in the cosmology of Mithraism, the influential cult of the ancient Persian god Mithras who was associated with the sun, representing the "soul's passage through the galactic gateways that open during eras of galactic alignment."

It seems hardly coincidental that Ophiuchus, which is an occulted thirteenth zodiacal sign symbolizing, among other things, DNA in the form of a coiling snake held by a figure similar to the Greek Laocoön, is located (from Earth's perspective) almost on top of Galactic Center. Here, in ancient symbolism, we find a direct link between the creational torsion waves emitted by Galactic Center and the DNA molecule to which, by all indications, they give rise.

Another amazing piece of evidence linking DNA to Galactic Center also came in 2006 when SPACE.com reported on a new study revealing that magnetic "forces at the center of the galaxy [had] twisted a nebula into the shape of DNA." This was the first time the double helix shape had been photographed in space.

"Nobody has ever seen anything like that before in the cosmic realm," commented the study's lead author, UCLA's Mark Morris. "Most nebulae are either spiral galaxies full of stars or formless amorphous conglomerations of dust and gas—space weather. What we see indicates a high degree of order."

"The DNA nebula is about 80 light-years long," reported journalist Bjorn Carey, and "about 300 light-years from the supermassive black hole at the center of the Milky Way." Carey goes on to explain that the "recipe for a DNA nebula is strict but simple. It requires a strong magnetic field, a rotating body, and a nebulous cloud of material positioned just right. Massive central black holes are the best sources for both the strong magnetic field and rotating body, and since most large galaxies have them, Morris expects DNA-like nebula may be common throughout the universe."

Long before the development of modern-day science, in ancient times the Vedics were well aware of the meta-genetic connection between Galactic Center and DNA, as well as the many cycles of time ending around 2012, employing the term *somvarta* to describe the intelligent waves of Core energy responsible for the spontaneous evolution of species.

Another ancient concept, the Golden Mean, precisely defines the mathematical relationship between "above" and "below." The DNA molecule is structured minutely on Phi or the Golden Mean, measuring 34 x 21 angstroms for each full helical spiral. Concordantly, the average mean orbit of each of the planets moving away from the sun is also a Fibonacci sequence that translates to almost exactly 1.618.

Wilcock is quick to bring the discussion of the "DNA wave," that is in a profound sense "spoken" into being by Galactic Center, back to the Shift in human consciousness that currently is being fostered meta-genetically on a macrocosmic scale: "DNA itself is a resonantly-tuned antenna for the 'consciousness field,' and our entire Solar System is experiencing an energetic charge-up that our own fossil record suggests will have a spontaneous evolutionary impact on our DNA ... triggering the 'Ascension' or 'Golden Age' so many prophetic traditions have spoken of."

10
Unconditional Love, Torsion Energy & Human Evolution

A second point related to human evolution that needs emphasizing is that the Shift now occurring is related directly to an increase in cosmic consciousness based in unconditional love. The Shift may be visualized as simultaneously evolution, revolution and, to borrow a term from the Weinholds, "LOVEvolution. "

For some, the current global blossoming of consciousness is viewed as a natural process of human evolution. To others, this phenomenon appears more radical, a spontaneous genetic leap forward. Still others believe that this step is merely the bringing forth of what has existed always as a human potential: a revolution back in the direction of wholeness and integration. I trust by now the reader understands that these ways of envisioning our species' present evolutionary phase are by no means mutually exclusive.

As the term *LOVEvolution* suggests, many believe that the dawning Age of Light or Age of Consciousness defines itself in relation to our capacity for unconditional love, our ability to transcend enemy patterning and victim consciousness while adopting unity consciousness that sees divinity in all things. From this standpoint, it might be said humans are evolving into a "biologically conscious" species capable of holding and sharing the light of unconditional love. As we will see momentarily, modern scientific research indeed supports the notion that the emotion of love is the key to true healing as well as conscious evolution.

It has been said that long ago, the ancient Maya conceptualized this next evolutionary stage occurring in the years leading up to 2013 as Mastery of Intention. Mastery of Intention appears to be another way of envisioning what I have been calling conscious personal mastery, which is achieved through activation of an embodied unity consciousness capable of infusing biology itself with new structures and possibilities quite outside the box of even much of today's "advanced" thinking about human bio-spiritual potential.

Conscious personal mastery is achieved through activation of an embodied unity consciousness capable of infusing biology itself with new structures and possibilities quite outside the box of even much of today's "advanced" thinking about human bio-spiritual potential.

According to Joseph Chilton Pearce in *The Biology of Transcendence: A Blueprint of the Human Spirit,* "Transcendence is our biological imperative, a state we have been moving toward for millennia." The title of another intriguing study by Pearce neatly summarizes the name of the endgame we now are playing: *Evolution's End: Claiming the Potential of Our Intelligence.*

How does one master intention in order to claim this potential? How does one consciously evolve bodily into "transcendence through immanence"? In other words, how can we best foster conscious personal mastery to facilitate our own metamorphosis, our own transmutation into a higher way of being?

The previously cited research by Bruce Lipton proves that consciousness can reprogram DNA. Our discussion of this topic in the last chapter centered on galactic consciousness, torsion waves spurring human evolution as Earth moves into a denser, brighter and "more conscious" area of the Photon Band. But what role does an individual's consciousness play in this cosmic drama of becoming?

Clairvoyance is unnecessary to see that human consciousness is expanding at a tremendous rate. A trip to the local bookstore demonstrates that consciousness, along with associated terms such as "intention" and "manifestation," has become a cultural buzz word.

Even having been steeped for many years in mainstream academe, when I began writing my seriocomic novel series *Beginner's Luke*—which played a big role in my own healing and, for all its parodic elements, shares the same core philosophies as this book—in earnest during my illness back in 1999, there was only one central theme that truly inspired me: the role played by consciousness in creating and re-creating our world, individually and collectively.

It must be emphasized that the exponentially increasing focus on consciousness we are witnessing is not merely a "new age" phenomenon, as some in the old guard have tried to paint it. Popular bestsellers such as Michael Talbot's *The Holographic Universe* and Larry Dossey's *Reinventing Medicine* make a clear and compelling case that science is beginning to admit the ancient hermetic principle (which has been adopted by many postmodern thinkers, including myself) that Mind is reality's primary building block.

On this subject, renowned psychiatrist Stanislov Grof has written that "modern consciousness research reveals that our psyches have no real or absolute boundaries; on the contrary, we are part of an infinite field of consciousness that encompasses all there is—beyond space-time and into realities we have yet to explore."

Such an expansive view of consciousness also informs Leonard Horowitz's review of the science of quantum holography, where he reminds us not only that a unified field of consciousness exists, but also that it "may be explained as emerging from a previously overlooked physical vacuum or energy matrix." From a human perspective, based on mounting evidence I barely have touched on so far, this nonlocal energy field functions through subquantum connections between DNA and universal creative consciousness or torsion energy.

On much the same wavelength, in *The Divine Matrix: Bridging Time, Space, Miracles, and Belief,* Gregg Braden examines the vast implications of three genetic experiments conducted between 1992 and 2000 that "shatter" the old materialistic paradigm on which traditional Newtonian science, and the resulting mechanical view of the body as a machine isolated from mind and spirit, are based.

The first of these experiments was Gariaev and Poponin's discovery of the "DNA Phantom Effect," which proves, to quote Braden, that: 1) "A type of energy exists that has previously gone unrecognized"; and 2) "Cells/DNA influence matter through this form of energy." (See the discussion of this mind-blowing Effect in Chapter Twelve.)

The second experiment, reported in the journal *Advances*, was performed by the United States Army in the tradition of similar experiments conducted by Cleve Backster. The Army's experiment clearly demonstrated that the connection between DNA and emotion continues intact following physical separation between a person's DNA (sampled from inside the person's mouth) and the actual person experiencing the emotion.

In Braden's words, this experiment suggests that 1) "A previously unrecognized form of energy exists between living tissues"; 2) "Cells and DNA communicate through this field of energy"; 3) "Human emotion has a direct influence on living DNA"; and 4) "Distance appears to be of no consequence with regard to the effect."

The third and final experiment cited by Braden is the extraordinary research of cell biologist Glen Rein on the impact of coherent human emotion on DNA. Here, DNA from human cells was isolated in a glass beaker and then analyzed (chemically and visually) in order to determine the impact of clearly sustained emotions, negative and positive, on genetic material as well as expression. According to Dr. Rein, "These experiments revealed that different intentions produced different effects on the DNA molecule causing it to either wind or unwind."

Rein and his colleagues discovered that anger, fear and similar emotions have the power to "wind" the DNA molecule, literally compressing it in a potentially harmful fashion. On the other hand, emotions such as joy, gratitude and love "unwind" or decompress DNA exposed to them, making DNA stronger and healthier. Similar conclusions have been reached by other researchers including Dan Winter, but to the best of my knowledge, Rein was the first to offer convincing and coherently articulated experimental proof of DNA's ability to expand and contract when emotionally prompted.

According to Rein's data, *it even may be possible for positive emotions rooted in love to revive or resurrect DNA apparently destroyed by negative emotions*—an astounding phenomenon with truly enormous implications. Lest the skeptical reader dismiss this possibility, let us ponder the related implications of Gariaev's assertion that "research in wave-genetics ... reveals potential applications with significant prospects for solving issues regarding the aging process and thus increasing life expectancy. This view is solidly grounded in [experimental] evidence."

Rein's research makes a compelling connection between life-giving torsion energy and uplifting emotions, particularly unconditional love, indicating that love promotes healing and also literally may propel evolution. Only the love-based emotions stimulate DNA to become healthier and thus more capable of interacting productively with environmental stimuli.

Rein's findings on the impact of coherent emotion on DNA can help us answer the critical question of how to participate consciously in our own evolution in a very specific manner. His research makes a compelling connection between life-giving torsion energy and uplifting emotions, particularly unconditional love, the most "coherent" of all emotions, indicating that love promotes healing and also literally may propel evolution (Free).

Only the love-based emotions stimulate DNA to unwind and become healthier and thus more capable of interacting productively with environmental stimuli. Hatred, depression, boredom and the like cause DNA to wind, destroying the viability of genetic information necessary for healing as well as evolution.

In keeping with Rein's research, Barbara Marciniak in *Path of Empowerment* writes that "genuine feelings of love and appreciation for your body convey a positive message containing essential life-sustaining signals that result in excellent health."

In direct contrast, maintaining "feelings of doom and despair, loneliness, helplessness, denial, anger, resentment, jealousy, greed, and fear conveys a negative message that promotes discord within the physical functions of the body." Marciniak concludes that the "ability to both give and receive love ... holds the true key to healing because it is the most life-sustaining and affirming form of emotional expression."

Rein's brilliant research, supported by Marciniak's inspired keys for surviving and thriving in a chaotic world in the process of transformation, indicates that *the single most important factor in our personal evolution is our commitment to open ourselves to our own healing by giving and receiving the primary torsion wave known as unconditional love.*

Appropriately, Wilcock's evolutionary model, founded in *The Law of One*, is crystal clear on the point that Earth and humans are evolving from a logos anchored in the third dimension to an existence rooted in fourth-dimensional, heart-based consciousness.

In *The Biology of Transcendence*, Pearce advances a hypothesis supporting this radical assertion based on the little-known fact that humans actually possess four neural centers in addition to the brain. One of these, currently in a state of development, is the "brain" located in and around the heart. Not surprisingly, the fourth or "heart" chakra, which in the Regenetics model is linked to the fourth dimension, known as the fourth "density" in *The Law of One*, typically is associated with Christ consciousness or unconditional love.

It is crucial that we understand the evolutionary engine behind this momentous developmental stage for our species, unconditional love, not as a weak abstraction, but as an omnipotent creational force of torsion energy that birthed—as it still is birthing—everything in the multiverse, including ourselves.

Unconditional love is named aptly because the creative principle of love places no conditions on its creations, allowing for the exercise of free will in the upward karmic spiral of evolving human consciousness. The Bible sums up this foundational concept in three words: *God is love*. For the ancient Egyptians and Maya, to cite but two examples, such infinite love associated with the life-giving feminine principle emanates from Galactic Center, also called the Central or Healing Sun.

Today this Core of our galaxy is thought by most scientists to be a black hole of massive proportions: the equivalent of 4 million of our suns, or more. For decades it was believed black holes annihilate whatever falls into them. Not too long ago, however, physicist Stephen Hawking performed an abrupt about-face when he admitted that black holes may indeed permit information inside them to escape. If we understand "information" to include hyperdimensional torsion-wave codes that create and modify life such as those that find expression meta-genetically via transposons in potential DNA, life indeed may originate, as whole civilizations of ancients claimed, from black holes.

Alternatively, it is possible that Galactic Center contains not only a black hole, but also a "white hole." In the words of investigative mythologist William Henry, "Of all the high-energy photons beamed at us by the Core, probably none are more puzzling than those emitted in gamma ray bursts. Astrophysicists speculate these bursts are coming from a white hole, a 'cosmic gusher' of matter and energy ... [W]hatever a black hole can devour, a *white hole* can spit out. These white holes precisely conform to the image the ancients held of the center of our galaxy."

The crisis, as well as the opportunity, of our time is to surrender the controlling aspects of our ego and its conditioned fear mechanisms to the primary torsion energy of unconditional love that is seeking to evolve us and is calling us as a species home.

In Chinese the word for "crisis" can be interpreted as "dangerous opportunity." "Within a larger framework of reality," writes Marciniak, "a crisis can be thought of as a meeting of minds at the crossroads of opportunity—a juncture where you recognize exactly where you are and consciously choose the best possible outcome for where you are going." The crisis, as well as the opportunity, of our time is to surrender the controlling aspects of our ego and its conditioned fear mechanisms to the primary torsion energy of unconditional love that is seeking to evolve us and is calling us as a species home.

This "home" may be simply a state of awareness that transcends duality and consciously exists in a multidimensional continuum. Wilcock sees "returning home" as a dimensional Shift referred to in the Bible as *Ascension* and in *The Law of One* as the "Harvest," in both cases envisioned as a spontaneous metamorphosis or transmutation involving consciousness as well as biology similar to what happened to Jesus in the Resurrection.

"There is a parallel in the Shroud of Turin," Wilcock notes, "where certain researchers have found that Jesus' body burned a complete three-dimensional image of itself into the cloth." Through experimentation it was determined that "such a burn could only be caused by an instantaneous blast at a very high temperature, 'zapping' the cloth like an X-ray."

Others also visualize returning home in terms of a radical Shift. Barbara Hand Clow has remarked that, in the final analysis, all dilemmas are perceptual—which implies that all solutions are perceptual as well. This sentiment is echoed by Judith Bluestone Polich in *Return of the Children of Light*, where humanity is described as standing on the brink of a collective perceptual "awakening. As the cosmic cycles of time are telling us, it is the time for a major turn upon the spiral path of evolving human consciousness, when the light that has descended into matter begins the ascent back to its origin."

The preceding quote also suggests that the home to which we have been referring may indicate some other place entirely. The 2012 alignment of the December solstice sun with Galactic Center creates what some ancients called the Black Road (Figure 6). It is conceivable we are meant to follow the Black Road home to our transdimensional Source, where we experience a state of being that altogether transcends this holographic reality composed of various "frequency domains," dimensions, or densities.

After all, a white hole is basically a black hole reversed. The two are thought to meet at their small ends like a pair of funnels (Figure 8). Mathematically, it should be possible to enter a black hole and emerge from a white hole in a completely new universe.[14]

---

[14] Something very similar happens to Jodie Foster's character in the movie *Contact*, based on the novel by astronomer Carl Sagan. The concept of the Black Road also finds support in physicist Lee Smolin's theory of "cosmological natural selection," mentioned in the previous chapter.

11
Becoming Light

It also must be stressed that the personal and collective evolution of consciousness and physiology we are embarking on presently as a species involves a positive genetic (trans)mutation that empowers humans to expand our multidimensional awareness while, effectively, becoming light.

Everything is energy. Einstein expressed an understanding of this truth in his famous theorem $E=MC^2$, which established the interchangeability of matter and energy. Concerning matter, Einstein once remarked, "we have been all wrong. What we have called matter is energy, whose vibration has been so lowered as to be perceptible to the senses. There is no matter."

Arguably, this truth that now has been validated even further by the quantum sciences was known to the ancient Hindus when they employed the term *maya*, meaning the illusion often mistaken for reality. Without a doubt, vanguard Russian scientists such as Kozyrev, Gariaev and Poponin who have studied torsion waves understand that energy (including so-called matter) is consciousness, and vice versa.

The notion that everything is energy or consciousness directly applies to human biology. The materialistic, "Era I" view of the body as a machine that may run on energy but is somehow distinguishable from it, is fast giving way to undeniable evidence that we, too, at our most fundamental level, are manifestations of conscious energy.

The holographic model, to reiterate, views the ostensibly physical universe in terms of intersecting electromagnetic frequencies that, in effect, project the staggering illusions we think of as the world ... and ourselves. "When two waves [for example, sound and light] come together they interact with each other producing [a hologram]," writes Horowitz. "Information is processed and cell structures are organized by these forces including the structure and standing waves created by DNA."

The notion that everything is energy or consciousness directly applies to human biology. The materialistic, "Era I" view of the body as a machine that may run on energy but is somehow distinguishable from it, is fast giving way to undeniable evidence that we, too, are conscious energy.

By now it should not strike the reader as "coming from left field" to learn that in *Vibrational Medicine*, Dr. Richard Gerber concludes that matter, including human cells, is actually "frozen light." Horowitz reaches precisely the same conclusion, bluntly stating that humans are "crystallized or precipitated light." This assertion is consistent not only with torsion-wave research and the holographic model, but also with the findings of more mainstream physics, which has demonstrated that light (much like DNA) is capable of both carrying and remembering data.

An insightful perspective on how the human body operates via—and even is composed of—light is provided by Dr. David Jernigan in an article entitled "Illuminated Physiology and the Medical Uses of Light," where the author explains,

Top researchers believe that our thoughts cause the mind to set up a morphogenetic field, which in turn fuels bio-holographic ... projections in the heart. These projections in the heart use biophotonic (laser-like) coherent emissions to transmit information and control inputs to the DNA and from the DNA to the entire crystalline matrix to support the thought command. In theory, heart-generated light traveling through the liquid crystalline matrix "optical fibers" of the body can produce "supercontinuum light," thereby maintaining its coherence and resulting in the multi-system wide effects seen when one biophotonic emission frequency from the heart is sent through the body's crystalline matrix ... [DNA contains] photo-receivers and photo-transmitters, and it may be that DNA is where the coherent signal is split into supercontinuum light to produce the "super-biohologram" that is the human body. It would seem that our thoughts are commands to the heart. The heart photonically imprints the DNA with the information to make the thought command come true.

The concept of light as information, or "light in formation," is an old one that for centuries has found expression in various types of sacred geometry. Others have suggested that angels, often depicted as divine messengers, are really *angles* or rays of light that convey information (typically experienced as inspiration and often centered in the heart) from a celestial source.

This expanded conception of light as a form of consciousness underpins, for example, the Toltec worldview of Don Miguel Ruiz, medical doctor, shaman and bestselling author of *The Four Agreements*, whose cosmology includes the Photon Band as a connector between Galactic Center and our sun. With the latter being Earth's primary source of light or information, the reverence for the sun in virtually all pre-industrialized cultures appears not naïve, but an informed and deliberate focus on humanity's local source of universal creative consciousness.

"DNA (deoxyribonucleic acid, the basic substance, which takes the form of a double helix, found in the nucleus of every cell which is associated with the transmission of genetic information) is a specific vibration of light that comes from the sun and becomes matter," observes Dr. Ruiz in *Beyond Fear: A Toltec Guide to Freedom and Joy*. "Every kind of life on Planet Earth, from ... stones to humans, has a specific vibration from light that comes from the sun. Each plant, animal, virus and bacterium has a specific ray of light ... condensed by Mother Earth and the information carried in the light becomes matter."

> ***Human biology may be considered electromagnetic at the level of its manifestation from the torsion life-wave that sustains it.***

New neurological research indicates that humans' tremendous brainpower, even operating below ten percent of our capacity, results not just from biochemistry, but from the brain's impressive ability to function as a holographic data storage and retrieval system (a "hard drive") that employs different light angles to read information ("software").

This implies, as noted, that the brain is a sophisticated holographic biocomputer that operates through electromagnetic frequencies. Not surprisingly, DNA has been shown to function very similarly. Human biology thus may be considered electromagnetic at the level of its manifestation from the torsion life-wave that sustains it. As Deepak

Chopra has observed, human cells, far from being merely functional vessels, are in actuality electromagnetic fields of possibility and potential.

Human bioenergetic frequencies can be identified clearly in the aura. As detailed in Part I, researchers now generally agree that humans possess a detectable aura. Kirlian photography has captured this iridescent halo around the body for decades. In 1972 biophysicist Richard Alan Miller developed a field theory to explain the aura. More recently, Dr. Valerie Hunt, UCLA professor and author of *Infinite Mind*, actually measured the human aura with an EEG machine. Early in the 20th century, it was theorized that the aura comprises various frequency bands known as auric fields, and that these govern distinct aspects of human biology, psychology, and spirituality.

The electromagnetic fields can be thought of as a geometric matrix, a "Jacob's ladder" that allows access to increasingly subtle frequency domains. This unfolding of perception to the full range represented by the electromagnetic fields and corresponding chakras is what it means to become "multidimensional."

Each of these so-called electromagnetic fields also corresponds to, and interfaces with, a specific dimension. The third field, for instance, is keyed to the third dimension. The electromagnetic fields can be thought of as a geometric matrix, a "Jacob's ladder" that allows access to increasingly subtle frequency domains.

This unfolding of perception to the full range represented by the electromagnetic fields and corresponding chakras is what it means to become "multidimensional." It is believed by Wilcock, myself and others that eight dimensions (not counting the transdimension of Source) are available to human perception at our present evolutionary stage, which means that operating in the first three dimensions, as most people do, humans currently only access just over a third of "reality." This does not even take into account the probable existence of multiple parallel realities.

According to Braden in *The God Code,* also noted, the ancient Hebrew four-letter name for God is secretly code for DNA based on the genetic code's chemical composition. "Applying this discovery to the language of life," writes Braden, "the familiar elements of hydrogen, nitrogen, oxygen, and carbon that form our DNA may now be replaced with key letters ... In so doing, the code of life is transformed into the words of a timeless message [that] reads: *'God/Eternal within the body.'"*

If God is indeed in the body—and consciousness and physiology, from an evolutionary perspective, are linked inextricably—we must acknowledge that *divine consciousness is available in and through physicality.*

One highly intriguing aspect of DNA is that most people utilize only about ten—some say as little as two or three—percent of it. As previously remarked, the other ninety percent or more has been dismissed by mainstream genetic science as "junk."

Interestingly, the fact that we use at best ten percent of our DNA correlates to the fact that we use at most ten percent of our brain. Still more provocative is that, according to one scientific model, String theory, less than ten percent of the matter in the universe is visible. The other ninety percent or so is sometimes called "dark matter" and, given the nonlocal quality of the torsion energy at its base, very well may reside in other dimensions.

Could "junk" DNA actually have biologically transformative potential awaiting activation? Could it somehow activate the unused portion of the human brain? Could this brain activation succeed in opening our godlike perceptual faculties, allowing us to climb the "multidimensional ladder" of our electromagnetic fields and experience the invisible ninety percent of the universe? Following the time-honored wisdom of "As above, so below," could reports that these perceptual faculties indeed are emerging in many people, especially today's extraordinarily gifted children, have anything to do with an increase in torsion energy in the form of superluminal light emanating from Galactic Center?

Many believe the answer to all these questions is an emphatic *yes*. According to William Henry, physicists "have established that a vast cosmic ocean of quintessence ... invisible to our telescopes ... surrounds the visible galaxies. If they are right, this 'dark matter' ... that composes [what] we ... see 'out there' is also 'in here' ... This implies that 9/10 of ourselves is also unknown."

"It seems clear that our hidden human potential is now unfolding," observes Judith Polich. "The higher order of consciousness that represents our next leap in human evolution is now being activated, and we are evolving into a higher frequency of light [...] Moreover, it is possible that the hidden potential lies within the very fabric of our DNA."

Polich quotes evolutionary biologist Elisabet Sahtouris, who observes that "the history of evolution has repeatedly demonstrated that DNA is capable of rearranging itself intelligently in response to changing environmental conditions. Therefore, some types of mutation may not be random at all." In fact, according to Sahtouris, our "DNA may be capable of utilizing information and making conscious changes in its structure. That is, it may consciously direct the process of mutation, thus transforming a species."

In a wonderful article reprinted in *DNA Monthly* entitled "Living Systems in Evolution," Sahtouris elaborates on the current "changing of the guard" in our way of thinking about the nature of DNA:

> In molecular genetic biology this shift is supported by fifty years of research evidence that DNA reorganizes itself intelligently when organisms are environmentally stressed, and that the required information transfer often seems to obey some form of nonlocality rather than slower chemical or electromagnetic transmission. Rather than being the sources of variation and evolution, errors known to occur in DNA during reproduction and by cosmic radiation or other accidents are recognized at the molecular level and fixed by repair genes. Thus we see intelligence at work not only in higher brains, but in the lowliest of bacteria and cellular components. Clearly, we are moving toward a post-Darwinian era in evolution biology.

Braden, who began his career as a geological scientist, was one of the first from the scientific community to theorize, based largely on observable Earth changes, that our planet is experiencing some type of frequency increase that ultimately will activate the dormant potential of our DNA. As detailed in *Awakening to Zero Point*, this evolutionary activation, or "Collective Initiation," possibly relates to Earth's harmonic frequency, known as the Schumann resonance.

Although this is a scientifically controversial subject, and has yet to be substantiated adequately, a

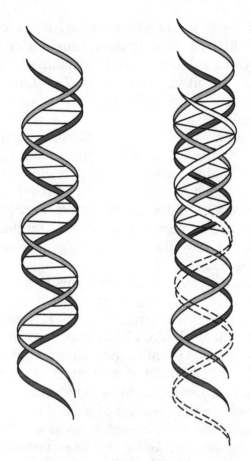

Figure 7: From Biology to Triology
The above illustration shows how a third strand of
DNA might interface with and modify the existing
double helix. Both as a possible biological reality and
as a metaphor for activating the latent intelligence in
potential DNA, the triple helix is genetically consistent
with the movement away from a binary or dualistic
"operating system" in favor of a trinitized or "trinary"
code capable of engendering an evolutionary Shift into
unity consciousness. Note how the geometry of three
strands naturally produces interlocking tetrahedral
shapes suggestive of molecular *merkabahs*.

number of researchers, myself included, still believe
that some, possibly higher-dimensional (torsion)

aspect of Earth's resonance indeed is increasing in keeping with the evolutionary timeline encoded in the Mayan calendar and explained in such compelling intuitive sources as *The Law of One.* Perhaps new data will facilitate our collective understanding.

Braden argues that Earth's hypothetical frequency increase, possibly linked to denser or brighter "light information" stemming from increased celestial activity, will result in new combinations of amino acids—in essence, new DNA. From a genetics perspective, this is tantamount to saying that a new life-form is emerging out of the human species.

As we already have seen, this is not nearly as odd as it at first may sound, given DNA's spectacular capacity for adaptive (trans)mutation. Recently, the phrase "quantum biology" has appeared in response (in some cases) to allegedly suppressed evidence suggesting that a third strand of DNA currently is activating in humans, forming what may be a "triple helix." I believe the creation of a third strand of DNA is a potential reality; however, even as a metaphor for evolving the latent intelligence of our existing DNA, the notion of a third helix has a certain conceptual value (Figure 7).

The generally accepted notion that DNA in its double-stranded form is a universal molecule actually finds little support when we examine the activity of DNA itself. As previously emphasized, DNA continuously is changing, mutating. To claim it is impossible for DNA to morph into another molecule represents an empiricist bias based on so-called immutable facts (that the DNA molecule has been around a long time in more or less the same configuration, for instance) which are called into question by, among other scientific discoveries, the Heisenberg Uncertainty Principle and the logic-defying realization that we create our reality by observing it.

Inevitably, the scientific community is coming around to an awareness of the mutable nature of what heretofore were considered nature's immutable laws. In the words of biologist Rupert Sheldrake in *The Presence of the Past*, the "assumption that the laws of nature are eternal" actually stems from early Christian influence on the scientific method that was developed in the 17th century. "Perhaps the laws of nature have actually evolved along with nature itself, and perhaps they are still evolving?" wonders Sheldrake. "Or perhaps they are not laws at all" but are "more like habits?"

In the theoretical case of radical genetic (trans)mutation from two to three strands of DNA, we still are looking at a single molecule of life. There is simply (or not so simply) the addition of a third strand requiring a reconfiguring of the manner in which nucleotide bases attach to the strands. This might occur through a large-scale genetic rearrangement as described by Colm Kelleher in his article "Retrotransposons as Engines of Human Bodily Transformation," a very helpful study which I touch on again in Chapter Thirteen.

On the subject of the emergence of a new genotype of human, Polich writes that "the codes to awakening our ancestral endowment—namely, our inner light—may lie hidden within the structure of our DNA." As we individually "begin to remember who we are, a new consciousness will emerge. [As] soon as this revisioning reaches a critical mass, it will trigger an evolutionary leap to a new human species—the long-awaited, quantumly endowed spiritual human known to ancient cultures as the child of light."

All living beings emit light. Anticipating the latest Russian research in wave-genetics, in the 1920s another Russian scientist, Alexander Gurvich, pioneered the concept of light frequency signaling via "mitogenetic rays" in human cells. Later in Germany, Marco Bischof published an influential text entitled *Biophotons: The Light in Our Cells.* By 1974 German biophysicist Fritz Albert Popp's biophoton theory had confirmed the basic mitogenetic hypothesis, demonstrating that DNA is the source of bioluminescence. Popp's theory, in turn, was corroborated by Herbert Froehlich and Nobel winner Ilya Prigogine.

In biology circles, more and more attention is being paid to a system of "biophoton light communication" that appears essential to many regulatory processes in living organisms. The cellular hologram equivalent of the nervous system, this intricate communication web that employs light for data transfer operates far more quickly than the nervous system and may be considered a real-time (parallel-processing) quantum biocomputer allowing for an unmediated electromagnetic interface with the individual's environment.[15]

Sheldrake's Morphic Resonance theory strongly suggests that cellular bioluminescence (which in humans ranges from ultraviolet to infrared) is both personal and transpersonal. In other words, not only is the individual human "networked" with DNA light emitters and receptors; it also appears our entire species is networked morphogenetically much

---

[15] This interface, I contend, initiates at the level of the electromagnetic fields, the "mind," and operates by way of the genetic sound-light translation mechanism described in Chapter Six. Rather than biophoton light communication, perhaps a more accurate name for this data transfer system, one that respects the primacy of sound at the ener-genetic level, would be "biophonon-photon communication."

like individual cells that form a larger biological entity: humanity.

This assertion has been substantiated by the Gariaev group, whose findings liken DNA not just to a holographic biocomputer, but to a "biological Internet" that links all human beings. Many native wisdom traditions are based on an equivalent understanding of the universe (human inhabitants included) as a single living being intelligently networked like a biological organism. A similar understanding of the ultimate systemic unity of reality is the absolute foundational concept behind *The Law of One.*

"Eastern teachings tell us that the living light is encoded in our form," writes Polich. "The ancient concept of the macrocosm as microcosm ... tells us that the greater divine [light] is reflected in the human body. Expressed in another manner, this means that ... the spiritual human ... has encoded within it a divine blueprint." This divine blueprint, which Polich refers to following the kabalistic tradition as the *Adam Kadmon,* also has been called the *lightbody.*

Throughout the centuries, the lightbody has been given many names. Other historical names for the lightbody include the "Diamond Body" and "Jade Body" (Taoism), the "Merkabah" (Kabala), the "Adamantine Body" (Tantra), the "Glorified Body" (Christianity), "Holy Flesh" (Catholicism), the "Superconductive Body" (Vedanta), the "Supracelestial Body" (Sufism), the "Radiant Body" (Neo-Platonism), the "Immortal Body" (Hermeticism), the "Body of Bliss" (Kriya Yoga), the "Perfect Body" (Mithraism), and the "Golden Body" (*The Emerald Tablets*).

According to author and researcher Sai Grafio, who contributed to the foregoing list, the Tibetan Buddhist tradition is the only ancient tradition that actually refers to our ascensional vehicle as the lightbody. It goes almost without saying that the building of any kind of body, "natural" or "supernatural," is accomplished by way of life's building block, DNA, in one form or another.

With great prescience in the middle of the 20ᵗʰ century, Indian philosopher and holy man Sri Aurobindo referred to the lightbody as the "Divine Body," explaining that it "is indeed as a result of our evolution that we arrive at the possibility of this transformation ... *Here a slow and tardy change need no longer be the law or manner of our evolution*; it will be only so to a greater or lesser extent so long as a mental ignorance clings and hampers our ascent" (my emphasis).

In the Bible, the lightbody is referred to as the "Spiritual Body" and the "Wedding Garment." The body's resurrection leading to its glorification is of paramount importance to the Catholic faith, which maintains detailed records of lightbody activations in its strict records used in the canonization process for saints. Perhaps the most famous biblical passage that appears to describe the transformation of the human body into a lightbody occurs in Saint Paul's epistle I Corinthians 15:39-44:

> All flesh is not the same flesh: but there is one kind of flesh of men, another of beasts, another of fishes, and another of birds.
>
> There are also celestial bodies, and bodies terrestrial: but the glory of the celestial is one, and the glory of the terrestrial is another.
>
> There is one glory of the sun, and another glory of the moon, and another glory of the stars: for one star differs from another star in glory.
>
> So also is the resurrection of the dead. It is sown in corruption; it is raised in incorruption.
>
> It is sown in dishonor; it is raised in glory; it is sown in weakness; it is raised in power.

It is sown a natural body; it is raised a spiritual body. There is a natural body, and there is a spiritual body.

James Redfield, Michael Murphy and Sylvia Timbers cite this passage in *God and the Evolving Universe: The Next Step in Personal Evolution,* a thoughtful and scholarly study of planetary evolution and human potential whose focus—surprising for a National Bestseller—turns out to be "luminous embodiment." In light of our current discussion, one passage from this illuminating book is worth quoting at length:

> [W]e can begin to picture bodily changes that might accompany the further development of our greater attributes, supposing: first, that esoteric accounts of bodily transformation, though frequently fanciful, reflect actual developments of physical structures as yet unrecognized by science; second, that supernormal capacities, like their normal counterparts, require distinctive types of supporting structure and process; and third, that we can extrapolate from physiological changes already revealed by modern research in imagining bodily developments required for high-level change.

Clearly, the lightbody is no starry-eyed, esoteric fantasy, but at the very least a biological possibility—one that, by various indications, gives rise to a radically new "spiritual" biochemistry and genetics that allow for the incarnation of the light of unity consciousness in every cell of the body.

It appears that the lightbody begins to express itself when dormant potential DNA codes are keyed by torsion energy, particularly *sound* and *intention* spiral standing waves of a hyperdimensional nature derived from the primary torsion energy of

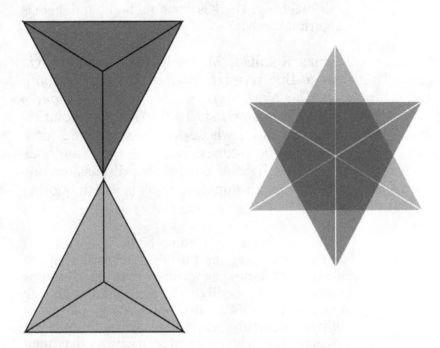

Figure 8: Tetrahedron & *Merkabah*
Note how from a three-dimensional perspective, the merkabah is an interface between a downward-pointing tetrahedron and an upward-pointing one, creating a molecular marriage between "above" and "below." It also is interesting to remark the similarity between this convergence and that of the theoretically funnel-shaped black and white holes approaching each other at Galactic Center.

unconditional love, such that our cells begin to recognize light as an energy source and metabolize it somewhat like plants do in photosynthesis.

I am referring specifically to the invisible (superluminal) torsion light emitted by the Healing Sun and transmitted to us from Galactic Center outside space-time by way of the Photon Band and our sun. This meta-genetic energy is a frequency or series of frequencies with linguistic characteristics that we can emulate microcosmically through linguistically based sound and intention, with or without machines, as I am endeavoring to explain and substantiate throughout this book.

Among other benefits, the cellular evolution prompted by this energy is capable of significantly increasing metabolism—a phenomenon regularly observed in clients of the Regenetics Method, and one that encourages detoxification, rehydration and, ultimately, rebuilding.[16]

This meta-genetic energy is a frequency or series of frequencies with linguistic characteristics that we can emulate microcosmically through linguistically based sound and intention, with or without machines.

In the profoundly transformational process of lightbody activation, which for many of us is likely to be experienced as a global event sometime around December 21, 2012, it would appear from the esoteric literature that the liquid crystals of cells evolve from primitive hexagonal structures observed in normal human tissues to what have been called "stellar" tetrahedrons. Needless to say, this structural change is extremely consequential for biological expression.

In such a scenario, the hydrogen bond angles of our water molecules literally may broaden and become interlocking tetrahedrons in order to hold more light or photonic energy that partially results from the enlarged hydrogen angles. These structures that form the liquid matrix for the new blood and

---

[16] On a related subject, Horowitz cites evidence that phototherapies (including sunlight) may "1) change [cell] membrane permeability; 2) increase cellular nutrient and mineral entry into the cell, and 3) facilitate release of toxins from the membrane and cell interior."

tissues are thought to resemble three-dimensional Stars of David or molecular *merkabahs* (Ascension Alchemy). Merkabah is an old kabalistic word meaning "vehicle of light" (Figures 7 and 8).

This helps explain the confusion that often has surrounded the merkabah, which sometimes has been taken to denote a type of spaceship. The odd truth is that the merkabah is a sort of spaceship—and that this spaceship is the individual's genetically activated lightbody.

According to Marciniak, the merkabah "represents the figure of the human being in its most unlimited state—the totally free human ... The lightbody is the body that holds the complete mutation of the species. It [can] juggle realities through the shifting of consciousness by intent" like changing channels on a television. In a similar vein, Tashíra Tachí-ren describes the merkabah as a "crystalline Light structure that allows you to pass through space, time, and dimensions, completely in your totality."

Whatever its formal geometry may be, and there are various competing theories on this subject, the lightbody results from and operates through embodied higher light or consciousness. It represents the natural result of a perceptual evolution into enlightenment, based in unity consciousness and unconditional love, that is, by many indications, occurring on a planetary level with gathering speed and power.

From a certain perspective, there is perhaps no such thing as ascension, only "descension" of the light of unity consciousness into physical form. The lightbody also may be, literally, the spaceship that allows us to travel the multi- and transdimensional Black Road through the stars ultimately all the way back to our Source.

In her own version of cosmological natural selection, Tachí-ren sees "going to Lightbody [as] only a part of a much larger process [in which] all planes and dimensions [merge] back into the Source for this universe, which then merges with other Source-systems, and so on, back to the One." However we define returning home, the lightbody— which itself appears to evolve into even higher expressions over time—is the vehicle that takes us there.

12
DNA Activation, Healing & Enlightenment

A variety of techniques exist to facilitate adaptation to new forms of thought and their corresponding biological structures or "thought-forms." These techniques are designed to be practiced co-creatively by individuals, even as the universe also meta-genetically prepares us for our bio-spiritual transformation into beings of greater and purer light.

Over the centuries, hundreds of modalities have been developed to assist in unfolding the spiritual human (the "Holy Grail") that exists as a genetic potential in everyone. Alchemy is an excellent example from antiquity. Etymologically, *alchemy* derives from the Arabic words *al* (the) and *khame* (blackness) and might be defined as the science of creating light out of darkness. As students of this discipline eventually discover, alchemy's real goal is not to turn lead to gold, but to transform human biology into a physiology of golden light. Stated a bit differently, alchemy's primary objective is enlightenment, or creation of the lightbody.

Today many more lightbody activation techniques are being made available, which itself signals a significant movement toward global bio-spiritual enlightenment. Leigh and I have been blessed to develop one such technique. The Regenetics Method employs specific combinations of sound and intention, which are differentiated aspects of the primary torsion energy of unconditional love, to stimulate the latent potential in DNA designed to facilitate the evolution of human beings into unity consciousness and its corresponding physiology of light.

DNA, as Braden points out in *The God Code*, is by its very sacredly encoded nature a unifying principle for humanity—one capable of bringing peace and harmony to the Twelve Tribes of a planetary population faced with a decision between succumbing to crisis or embracing opportunity.

"The discovery of God's name within our bodies shows us the benefit of merging different ways of knowing into a single understanding," he writes. "By crossing the traditional boundaries that define chemistry, language, history, and religion, we are shown the power of a larger, integrated worldview." That we can activate DNA to expand our worldview and, in the process, assist the cosmos in evolving a biology fit for unity consciousness is truly a divine gift at this crossroads in history when, to quote Barbara Marx Hubbard, we must "decide between conscious evolution, or extinction through misuse of our powers."

The Regenetics Method features four integrated DNA activations that, collectively, help to establish the ener-genetic template for unfoldment of the lightbody. The first of these, Potentiation, activates DNA to repattern the body's electromagnetic fields, resetting the human bioenergy blueprint to an infinity circuit based on the number 8 as detailed in Part I.

DNA has been compared to an antenna connecting humanity to Galactic Center or Source whose reception can be clouded by toxicity and trauma. Potentiation activates DNA to begin removing this toxicity and trauma, establishing a clearer connection through harmonic resonance with the unconditional love frequency emanating as hyperdimensional spiral standing waves of sound and light from the Healing Sun at Galactic Center. In other words, Potentiation attunes DNA to the primary torsion energy of Source.

> *Through harmonic resonance, Potentiation establishes a clearer connection with the unconditional love frequency emanating as hyperdimensional spiral standing waves of sound and light from the Healing Sun at Galactic Center. In other words, Potentiation attunes DNA to the primary torsion energy of Source.*

As mentioned, Leigh and I consider ourselves merely facilitators for the individual's own bio-spiritual unfoldment. We actually heal ourselves. This is a critically important point that typically is overlooked, but cannot be overemphasized.

In *The Law of One*, a philosophy highly consistent with the basic principles that underwrite the Regenetics Method, it is stated explicitly that healing "occurs when [the self] realizes, deep within itself, the Law of One; that is, that there is no disharmony, no imperfection; that all is complete and whole and perfect. Thus, the [Creator] within this [person] re-forms the illusion of body, mind, or spirit to a form congruent with the Law of One. The healer acts as energizer or catalyst for this completely individual process."

As William Henry reminds us, the "Healing Sun rises from within us when we place ourselves in balance with its energies ... By conceiving of these healing energies ... we can tune into them. What we can conceive we can achieve." I would add that we place ourselves in balance with the Healing Sun not by merely looking inside ourselves, but first and foremost by adopting an internal attitude of unconditional love.

My illness, which had affinities to chronic fatigue syndrome (CFS or CFIDS), multiple chemical sensitivity (MCS) and fibromyalgia, was precipitated by toxicity and trauma from a series of hepatitis and yellow fever vaccines I received in the spring of 1995. The reader will recall from our discussion of this topic in Part I that vaccines potentially damage, and even alter, human genetics. Moreover, and more disturbing still, based on observation and kinesiological testing, such damage and alteration can be inherited by children who never receive vaccinations physically.

In the words of vaccine whistleblower Neil Z. Miller in *Immunization: Theory vs. Reality*, on top of deliberately planned additives such as aluminum, formaldehyde and mercury, vaccines also may contain "unanticipated matter. For example, during serial passage of the virus through animal cells, animal RNA and DNA—foreign genetic material—is transferred from one host to another. Because this biological matter is injected directly into the body, researchers say it can change our genetic makeup."

In my case, in the aftermath of my debacle with vaccines, after years of intense suffering and trying one expensive (and mostly ineffective) therapy after another, my turning point came when I realized that if I somehow could find a way to reset myself at the genetic level, my thirty or so debilitating symptoms eventually would go away.

To recap, I found myself on this path after reading an alarming book by Horowitz called *Emerging Viruses.* Basing his claims on meticulous research, Horowitz demonstrates that vaccines are a principal cause of a variety of autoimmune diseases, including AIDS. He further exposes what is in essence covert biowarfare conducted by the medical establishment against an unsuspecting population in *Healing Codes for the Biological Apocalypse*, where a main theme is the use of sound to heal the physical body by restoring it to genetic integrity.

According to Horowitz, the specific sounds used to restore genetic integrity derive from the ancient Solfeggio scale. This primordial six-note scale, which was dubiously "lost" by the Roman Church some centuries ago, was rediscovered recently by coauthor Dr. Joseph Puleo as described in *Healing Codes*.

Use of the Solfeggio scale in alternative healing has become increasingly popular since the publication of *Healing Codes*. In a questionable move, Horowitz has taken the liberty of extrapolating an additional three notes from the scale's intervals, forming a nine-note scale—which I contend is neither functionally necessary nor historically warranted.

All four DNA activations of the Regenetics Method employ notes from the original Solfeggio scale, which some scholars believe to be the sacred set of six notes employed by the Creator to fashion the world in as many days.

Be that as it may, here it is simply necessary to point out that all four DNA activations of the Regenetics Method employ notes from the original Solfeggio scale, which some scholars believe to be the sacred set of six notes employed by the Creator to fashion the world in as many days. Specifically, Potentiation employs the note "Mi," a frequency (528 Hz.) that has been used by leading-edge molecular biologists to repair genetic defects.

Having discovered Horowitz and Puleo, I was then fortunate enough to stumble on another book that greatly expanded my awareness of the immense scope of human potential: *The Cosmic Serpent: DNA and the Origins of Knowledge*, by French anthropologist Jeremy Narby. *The Cosmic Serpent* is an exploration of DNA from a shamanic perspective that describes how sound can be used to stimulate a genetic self-repair mechanism as detailed in Part I.

Intrigued (and desperate) enough to delve into this subject, I read in an eye-opening article by Bärbel Mohr that the power of sound to activate DNA recently had been documented by the Gariaev group in Russia. Dr. Gariaev and his brilliant team of geneticists and linguists proved that damaged DNA can be healed without gene splicing or other physical manipulation by merely immersing it in radio and light waves keyed to human language frequencies.

According to Gariaev's groundbreaking research in wave-genetics, DNA not only assembles proteins through RNA transcription, but also stores and communicates data in a decidedly *linguistic* fashion. His team found that the genetic code in potential DNA observes, for practical purposes, the same foundational rules as verbal communication.

In order to prove this, the grammar, syntax and semantics of language and DNA were analyzed together. It was discovered that potential DNA's alkaline sequences closely mirror linguistic communication rules. This establishes that the many human languages could not have appeared randomly, as is commonly believed and taught, but point to our essentially similar genetic structure. Supported by Braden's discovery that the ancient Hebrew name for God is code for DNA, Gariaev's findings offer stunning scientific corroboration that in the beginning was the Word!

Gariaev and his team also studied DNA's uncommon electromagnetic behavior. They concluded that chromosomes *in vivo* function exactly like holographic biocomputers powered by DNA's own lasing system. The Gariaev team modulated certain linguistic frequencies onto a laser. With this technology of linguistically modulated codes translated into radio and light waves, they not only were able to heal mutated chromosomes—they also regrew endocrine glands in animals, stimulated regrowth of new adult teeth in humans, and even successfully altered genetic expression.

Amazingly, they obtained results similar to those documented by Dr. Yu Dzang Kangeng, who was the first to employ torsion energy (although in a manner quite different from Gariaev's) to map DNA sequences between organisms. Using radio and light waves keyed to language frequencies to rewrite DNA, and dispensing altogether with gene splicing, *Gariaev's team literally transformed frog embryos into healthy salamander embryos in the laboratory.*

In this manner, metamorphosis was achieved without any of the side effects encountered when manipulating isolated genes. The "random and risky nature of gene splicing has been sadly hidden from the public," warns Horowitz. "With gene therapy, researchers cannot definitively predict where on a [chromosome] the modified gene might land, raising a hazardous possibility of inadvertently disrupting other genetic expressions and cellular functions."

Compare this to the organic alteration of genetic expression that can be achieved by simply applying vibration and language (or sound and intention, or *words*) to DNA. Gariaev's historic experiment in embryogenesis underscores the immense power of wave-genetics, an area which, contrary to current molecular biology dogma and propaganda, has an obviously more primary influence on the origin of species than biochemistry.

In Gariaev's own words, wave-genetics "may strike the reader as a technology of another age, from the distant future. However, the discovery of the fundamental properties of living organisms is occurring today, and it is our task to research and explain these phenomena and bring them forward in service to humanity."

Masters such as Jesus and the Buddha have always insisted that our genetic code can be "potentiated" through language—thus the healing effects of prayer, hypnosis, affirmations, mantras, etc. Happily, Gariaev's pioneering research now has scientifically substantiated such phenomena. The more developed the individual healer's consciousness, the less need there is for a mechanical crutch.

Human consciousness, not technology, is the key to true healing and enlightenment. Forever relying on something outside ourselves—such as a technological device—to heal us, or make us "whole," is at best giving away our power, at worst worshiping a false god.

J. J. Hurtak emphatically argues this point in *The Keys of Enoch,* stating that technology "must not be offered up as our consolation, for ... it would be our desolation. [We] must clearly see the spiritual dialectic taking place between those who choose ... Light as the touchstone for evolution ... [and] those who choose [technology's] codes for containment of the flesh, retrogressive evolution within three-dimensional form, and the annihilation of the hope for spiritual liberation."[17]

---

[17] In a similar vein, the Weinholds have written of what they call the Path of Technology (as opposed to the Path of Nature), "We have nothing against technology per se, and believe that some devices are helpful as 'boosters.' When created by minds connected to heart and spirit, technology can become a useful, but not essential, spiritual tool for enhancing genuinely spiritual living. Technology lacking connection with heart and spirit becomes just another box

Astonishingly, Gariaev's research reveals even more far-reaching implications with respect to the unlimited healing power of human genetic consciousness. The Russian team found that wave-activated DNA can manipulate the space-time matrix, producing small electromagnetic wormholes of a subquantum nature as covered in Part I. These DNA-activated wormholes, whose energy signatures are similar to those of Einstein-Rosen bridges found near black holes, are connections between different areas in the multiverse through which data can be transmitted outside space-time.

Potential DNA—which regulates meta-genetic functions—magnetizes these nonlocal streams of information to itself and then forwards them to our consciousness. Grazyna Fosar and Franz Bludorf, authors of a summary of Gariaev's findings entitled *Vernetzte Intelligenz* ("Networked Intelligence") (summarized in English by Mohr), refer to this data exchange as *hypercommunication*, pointing out that it often is experienced as intuition or inspiration.

When hypercommunication occurs, according to Fosar and Bludorf, an extraordinary phenomenon can be observed in DNA. They relate how Gariaev irradiated a DNA sample with a laser until a typical wave pattern appeared on his monitor. When the DNA sample was extracted, its electromagnetic pattern remained, perfectly intact. Many control experiments established that the pattern still

---

from which it is necessary, at some point, to break free." When technology transforms into a path, "it encourages dependency because it implies that people need to rely on a technical intermediary to help them open their own gates of perception. Rather than using primordial tools grounded in nature to empower initiates on their return to Source, the Path of Technology can end up disempowering people and encouraging spiritual co-dependency."

emanated from the absent sample, whose energy field remained undisturbed in the vacuum chamber for up to thirty days, causing light to spiral all by itself following the shape of the physically removed double helix.

This nonlocal, light-bending torsion energy phenomenon has become famous in new science circles as the *DNA Phantom Effect*. It is theorized that torsion waves from beyond space-time continue to emerge from the activated wormholes even after the DNA is separated out.

Remote (i.e., distance) DNA activation thus is explained easily as a meta-genetic transfer of consciousness manifesting as torsion waves that stimulate a molecular rearrangement of transposons in potential DNA. In turn, potential DNA shifts the bioenergy fields, which then modify metabolic and replication functions in cells, facilitating healing (Figure 2). This same process of DNA activation that encourages healing, by inviting more torsion light into cells, by definition promotes *enlightenment*.

Remote DNA activation is explained easily as a meta-genetic transfer of consciousness manifesting as torsion waves that stimulate a molecular rearrangement of transposons in potential DNA. In turn, potential DNA shifts the bioenergy fields, which then modify metabolic and replication functions in cells, facilitating healing. The same process that encourages healing, by inviting more torsion light into cells, by definition promotes enlightenment.

"Most people tend to think that the DNA created the [phantom] energy field, and that the energy field is somehow just a 'shadow' of the DNA," writes David Wilcock, who proposes a fascinating reinterpretation of the DNA Phantom Effect:

"However, I believe that the wave actually ex before the DNA ... [The] only logical explanation that the phantom energy of DNA is actually the *creator* of DNA." Since this phantom spiritual energy pervades the galaxy, wherever the "materials that create life exist, the subtle, spiraling pressure currents of this energy will arrange the DNA molecule into existence."

Combined with much research of my own, my intuition led me to theorize that the correct combination of sounds, intentionally geared to the body's bioenergy fields, could invite an influx of torsion energy from Galactic Center capable of clearing vaccination toxicity and trauma at the genetic level and upgrading the human electromagnetic blueprint to higher harmonic functioning.

I came to view the electromagnetic fields as an individual's energetic—and multidimensional—blueprint that not only can be reset like a blown fuse, but also, in the process, significantly upgraded. Leigh and I named this approach to DNA activation the Regenetics Method after Potentiation Electromagnetic Repatterning resolved my chronic illness and we began to develop Articulation, Elucidation, and Transcension. The "fringe benefit" of Potentiation was that it also promoted enlightenment by initiating the first phase of lightbody activation.

13
Historical & Scientific Overview of Enlightenment

Many ancient traditions worldwide maintain that humans not only inherently possess the potential for fully incarnating light at the physiological level, but that some already have achieved it, and millions more will do so in the very era in which we live.

The historical literature "suggests that there are unusual physical, as well as psychological, consequences in humans to the attainment of the exalted state of mind known as enlightenment," writes biochemist Colm Kelleher. "These reported changes include, but are not limited to, sudden reversal of aging, emergence of a light body and observed bodily ascension." While many of these descriptions associate the lightbody with death, Kelleher makes it clear that a number of reports indicate that "transformation of the body can happen independently of death."

The path of physical transcendence or bio-spiritual enlightenment through lightbody activation was embraced as a reality in most of the ancient world. The death and resurrection of Christ and Osiris are two famous examples from the Near East.

In the Middle Ages, a group known as the Cathars from southern France claimed to possess the secret gospel of Jesus, referred to as the Gospel of Love, believed to contain linguistic keys for creating the lightbody. The existence of this text became known after the Roman Church began to torture the Cathars in one of its infamous Inquisitions—at which point the Gospel of Love mysteriously disappeared like the Solfeggio scale.[18]

It seems that the Gospel of Love was, among other things, a manual for creating the merkabah. The merkabah, according to William Henry, is the "light body vehicle of resurrection and ascension that is the foundation of Hebrew mysticism. These texts make it clear that the Mer-Ka-Ba is a vehicle of light that emerges from within the human body."

Henry points to the Resurrection, following which the doubting fingers of Thomas appear to enter Christ's transfigured, luminescent flesh, as one famous description of the completely activated lightbody. In *Terra Christa*, Ken Carey also examines Christ's embodiment following the Resurrection, from Mary's perspective when first seeing Him in the garden. In explanation of Christ's strange warning to Mary not to touch Him, Carey writes that if

> Mary had touched Jesus at this point, the current would have likely killed her. It would have destroyed the circuits of her body-temple. As much as Mary loved Jesus, she did not have any comprehension ... *how much Christ loved*. The power of Christ's love coming to focus in the resurrected body of Jesus at that moment was for all creatures ... Mary needed to move into that kind of love. Like all of us, she had to learn how to thoroughly fulfill her human loves, loving her neighbors as herself. Then she might begin to approach the hem of the Master's garment.

---

[18] Bizarrely, owing to the stringent evidence requirements for canonization of saints, the Catholic Church maintains some of the most detailed records of paranormal phenomena associated with the lightbody, including several instances of individuals levitating, flying, or bilocating.

Learning to love one's neighbors as oneself starts, of logical necessity, with learning to love oneself. This teaching is at the heart of conscious personal mastery, the accomplishment of which is capable of transmuting the human body into a lightbody—an evolutionary transfiguration that has everything to do with healing not just the self, but also the world. "*We* are the world," writes Joachim-Ernst Berendt. "We cannot change the world if we do not change ourselves first. Any other way would be absurd." Happily, as individuals, we can encourage and condense this transformational process through a variety of "consciousness technologies," including DNA activation.

In the previously cited article "Finding the Holy Grail," which was inspired partly by the Regenetics Method, Barry and Janae Weinhold make a similar claim, pointing out on the basis of decades of in-depth historical research that the "individual human body *is* the Holy Grail. It isn't something 'out there.' Like a tuning fork, the body can be tuned to different frequencies," including that of Source, through DNA activation.

"Many indigenous traditions of Mesoamerica believe that [Source] emits a frequency or tone known as 'Ge' that not only heals the body-mind-spirit but provides immortality," write the Weinholds, adding that the

> spiritual practices used in the ancient mystery schools of Egypt and Greece employed a variety of ... vibrational tools to attune people's DNA to Source. This caused the DNA to ring, sing or vibrate so that it resonated with the tone of *Ge*—the frequency of Galactic Center. This attunement activated a *San Graal* or Song Grail—a "love song in the blood"—creating a rainbow bridge that synchronized an initiate's consciousness with

Source. This love song energetically united Heaven (Galactic Center) with Earth (initiates), opening human hearts and pumping crystallized Ge-tuned blood through their bodies. From this perspective, "Ge-sus" is a Master Being sent from Tula or Galactic Center to help humanity attune its DNA to the frequency of Ge so that we can return to Source.

In a nearly identical vein, Horowitz writes that "DNA seems to be transmitting the equivalent of heavenly love songs. From this music, played through genetic equipment, variations in sacred geometric forms materialize in space."

The profound and numerous connections between the Egyptian Osiris and Jesus have been noted by generations of scholars. Among many other similarities, both are linked to the phoenix or heron, represented in hieroglyphics as coming from Galactic Center, also (as indicated in the above quote by the Weinholds) called *Tula*, carrying the key of life.

The ankh or key of life is of a decidedly vibrational nature and designed to be employed along with a type of inspired (and inspiring) speech known as the* Language of the Birds. *This powerful combination, properly performed, keys potential DNA to begin building the Holy Grail or lightbody.

In Egyptian hieroglyphics, this key appears as an ankh, which may have been a type of actual tuning fork for harmonizing with Galactic Center, or simply may have symbolized appropriate musical notes (such the Solfeggio scale) for producing this celestial harmonization. In either case, the ankh or key of life is of a decidedly vibrational nature and designed to be employed along with a type of inspired (and inspiring) speech known as the *Language of the*

Birds. Based on personal experience and professional observation, I believe that this powerful combination, properly performed, keys potential DNA to begin building the Holy Grail or lightbody.

The Language of the Birds, according to the Weinholds who cite Henry's research, "is a vowel-only phonetic code ... Genetic and linguistic research indicates that the five vowels correspond to the five letters used to represent DNA and RNA ... Initiates of the Language of the Birds who are able to speak or tone these vowels in certain ways know that these sounds permanently activate the DNA of all those who are able and willing to hear."

At the genetic level, such a radical activation, according to Kelleher's previously discussed research in transposons, likely occurs through a *transposition burst* involving the molecular rearrangement of perhaps thousands of genes. Consistent with Braden's hypothesis regarding dormant DNA activation as well as Wilcock's compelling model of spontaneous evolution that simultaneously transmutes both consciousness and biology, Kelleher insists that true enlightenment, in addition to being a mental state, appears to have *physical* consequences.

The "appearance of a light body as a result of attaining enlightenment ... could be described as the emergence of a new species in a single generation from humanity," he writes, adding that a "synchronized, non random transposition burst is the most simple molecular mechanism to account for the required new configuration." While pointing out based on the historical literature that lightbody creation appears to occur only "in humans who have attained spiritual mastery," Kelleher emphasizes the existence of "stages on the road to ... enlightenment [that are] experienced by a great number of ordinary people."

This supports the potential effectiveness of a step-by-step process, such as the Regenetics Method, to bio-spiritual enlightenment in which transposons are stimulated incrementally in preparatory phases—culminating in a "synchronized, non random transposition burst" when, and only when, the individual is prepared consciously to experience it.

"There are preparatory steps as individuals approach the point where they are open to transformation," writes Ken Carey in his classic *The Starseed Transmissions*, a central theme of which is the planned occurrence of individual and collective luminous embodiment during our lifetime. "But the actual transformation is not a sequential process. It is not a complicated ritual. It can occur in the twinkling of an eye. It only involves one step, one decision, one event."

The reader may recall that another of Carey's books, from which the epigraph to this book is taken, is entitled *Return of the Bird Tribes*. In one apocryphal text known as the *Pistis Sophia*, Jesus (considered a master of the Language of the Birds) discourses on the afterlife in terms that appear straight out of the Egyptian *Book of the Dead*. Henry calls this "the first lesson of the Mer-Ka-Ba mysticism." "You are to seek after the mysteries of the Light," Jesus is quoted as saying, "which purify the body and make it into refined light."

Later, Jesus describes the connection between our dimension and higher dimensions as operating "from within outwards"—a statement, according to Henry, that "refers to a transformation of consciousness that opens the door to other worlds." Similarly, in the Bible Jesus insists, *The Kingdom of Heaven is within*. Horowitz's research leads him to affirm the material truth of this assertion: "The bioacoustic and electromagnetic matrix through which the Holy Spirit flows is real. It's what animates your DNA [by transmitting] the Kingdom of Heaven to you, and through you, right now, on Earth as it is in Heaven."

Another important figure from the Mediterranean associated with the lightbody is the Egyptian Thoth, called Hermes Trismegistus by the Greeks and considered the father of alchemy. Thoth is credited with the enormously influential phrase "As above, so below." Of the many seemingly miraculous gifts he brought his people, arguably Thoth's most important legacy is the doctrine of inner or spiritual light that literally can metamorphose the human body into a physiology of divine radiance. Interestingly, Thoth also was revered as the creator of writing and his alchemical science of transformation, like that of Jesus, is associated with the Language of the Birds (Henry).

The lightbody is also a theme in the ancient mystical traditions of Central and South America. A variety of mythical figures exist similar to the Incan god-man Amanumuru, who according to legend walked through a portal called the Muru Doorway and returned via the Black Road to his true home among the stars.

In what is today Mexico, a figure known as Quetzalcoatl embodied the higher light of divinity. Judith Polich has called Quetzalcoatl "the Osiris of Mesoamerica" and sees him as symbolizing the bridging of duality, a being who "represents light in physical form ... freed from the confines of matter."

Readers of Daniel Pinchbeck's phenomenally bestselling *2012: The Return of Quetzalcoatl* may recall this strong admonishment purportedly from the Mesoamerican god delivered through the author: "Soon there will be a great change to your world ... You have just a few years yet remaining to prepare the vehicle for your higher self. Use them preciously ... Consciousness is technology—the only technology that exists ... The current transition is ... a return to origin. The original matrix of this new world reality is the ecstatic limitlessness of your own being."

It seems poetically fitting that Quetzalcoatl nearly always is depicted as a hybrid figure, a great serpent with bird wings. In my opinion, this is an obvious reference to the ability of DNA (symbolized by the serpent) to transform one into a birdlike, angelic being capable of *f/light*—which is an excellent definition of the merkabah, as well as a wonderful visual illustration of how the "terrestrial body" of humans as we know ourselves (the serpent that exists on the ground) becomes the "celestial body" of ascended humans (the angelic bird that thrives in the air).

Equally meaningful is that Quetzalcoatl is thought to have revitalized the great ceremonial center of Teotihuacán, believed by some archeologists to be Earth's interdimensional gateway to the legendary Tula, the celestial home of Quetzalcoatl— and perhaps other "messianic" herons or phoenixes who arrived in this dimension carrying the key of life, or the knowledge of how to inspire ("breathe life") by speaking or singing the bio-spiritual Language of the Birds.

Elucidation Triune Activation, the third DNA activation in the Regenetics Method, is designed to establish the ener-genetic precondition for lightbody expression. This is done by employing the Solfeggio scale in tandem with what Leigh and I understand to be a form of the Language of the Birds to activate, by way of the neocortex and triune brain, the brain's prefrontal cortex, sometimes called (curiously enough) the "bird brain" as distinguished from the new mammalian, mammalian and reptilian cortices.

Within the brain's "triune nature," to use a phrase coined by neuroscientist Paul MacLean, we easily can chart the cognitive evolution of our species from the dominance of the reptilian brain; to the development of the limbic, old mammalian or emotional-cognitive brain; to the ascendancy of the neocortex, verbal-intellectual or new mammalian brain. Today, in addition to this nested tripartite structure that reads like a three-stop roadmap through human history, we are beginning to see activity in our "fourth brain," the prefrontal lobes.

"Neuroscientists have a variety of viewpoints on this comparatively new portion of our neural system, which was once called 'the silent area' of the brain because its function was largely unknown and no activity was indicated there," explains Joseph Chilton Pearce in *The Biology of Transcendence*. "Paul MacLean considered the prefrontals a fourth evolutionary system, however, and called them the 'angel lobes,' attributing to them our 'higher human virtues' of love, compassion, empathy, and understanding."

I believe that MacLean was basically correct, and that our transformation into a more highly evolved human necessarily involves activation, to some extent, of the prefrontal cortex. In an article entitled "Enlightenment and the Brain" republished in *DNA Monthly*, neurophysicist Christian Opitz reports that Sri Bhagavan, developer of a popular energy healing technique known as *deeksha*, insists that

> activation of the Frontal Lobes is involved in God-realization. The experience of enlightenment, of non-separation, does not necessarily coincide with the experience of a living God-presence. In Sri Bhagavan's teaching more than the deactivation of the overactivity in the parietal lobes is necessary

to move from enlightenment to God-realization. He speaks about the activation of the frontal lobes as a necessary neurological change for God to come alive in the consciousness of a person. The frontal lobes are associated with the individual will. Many mystical traditions speak about the merging of the individual will with the will of God as both a doorway to and result of God-realization. This cannot happen, however, if the frontal lobes are underactive.

Sri Bhagavan is making an important, often overlooked distinction between the meditative calming of the parietal lobes during the so-called enlightenment of many Eastern mindfulness practices, and genuine bio-spiritual *Enlightenment*, which involves not just perceptual oneness (transformative as that may be) but *physiological illumination*—what Sri Bhagavan calls "God-realization." Moreover, such divine embodiment clearly is stated to involve activation of certain areas of the brain.

Here, it must be stressed that much of the information mainstream science has propagated about biology is misinformation and perhaps disinformation: certainly incorrect and maybe intentionally misleading. For example, in humans the brain and central nervous system are not composed entirely of organic matter; they also appear to contain up to ten percent of the following six precious metals: gold, iridium, osmium, palladium, platinum, and rhodium.[19]

This fact has remained largely hidden from the public at least in part because current technology simply is not designed to detect such metals that

---

[19] This special family of metals should not be confused with environmental non-monatomic heavy metals, such as mercury and lead, which are highly toxic to human cells.

normally exist in human physiology in multidimensional states. Kinesiological testing reveals these six metals energetically correspond to, and can be activated by, the six notes of the Solfeggio scale.

Renowned historian Sir Laurence Gardner, bestselling author of *Bloodline of the Holy Grail* and *Genesis of the Grail Kings*, offers evidence that these six extraordinary metals that form part of the prefrontal lobes and central nervous system not only exist in humans, but can be made to align their atomic spins so as to become fully hyperdimensional and thereby create an internal source of torsion energy that manifests as light.

Certain "areas of the brain can be stimulated to extend human consciousness beyond any imagining," writes Gardner when discussing the prized alchemical knowledge contained in the *Book of Thoth*. By activating our internal "chassis" of precious metals, we can help stimulate creation of the lightbody from within. This internal energy often is called kundalini, which according to Vedic teachings has the potential to unfold one's bio-spiritual enlightenment (the lightbody) when fully awakened.

Combined with a specific vowel-only phonetic code that medieval alchemists referred to as the Green Language and I have been calling the Language of the Birds, the six notes of the Solfeggio scale can be voiced, as they are in Elucidation, to activate the six precious metals that form part of the prefrontal cortex.

Combined with a vowel-only phonetic code that medieval alchemists referred to as the Green Language and I have been calling the Language of the Birds, the six notes of the Solfeggio scale can be voiced, as they are in Elucidation, to activate the six precious metals that form part of the prefrontal cortex. This appears to occur after the neocortex (which also contains these multidimensional metals) is stimulated, creating a "cortical resonance" in the triune structure that then harmonically activates the fourth brain. It is for this reason Leigh and I chose "Triune Activation" to characterize Elucidation.

I am well aware this explanation may strike some of my more "left-brain" readers as far-fetched, and that proof of this phenomenon would be welcomed by many. My explanation here is my sincere and best attempt, based on extensive research over a number of years, to account for the powerful experience of personal transformation that often occurs following Elucidation Triune Activation.

In my own case, I felt decidedly "altered" for months following Elucidation, as if my brain chemistry had shifted fundamentally and permanently. I was and am reminded of Opitz's line in his aforementioned article that "[d]opamine, the essential neurotransmitter for frontal lobe activity, is necessary for feelings of enchantment with life and bliss, often described as accompanying mystical union with God."

My working theory is that this third DNA activation initiates a progressive, monatomic "high-spin" effect that produces bioacoustic and bioelectric signals capable of starting the process of turning the body's liquid crystals from hexagons into interlocking tetrahedrons or merkabahs (Figure 8). Apparently, this phenomenon is akin to the use of kind words to metamorphose the shape of water molecules as documented in recent Japanese studies conducted by Masaru Emoto.

Dr. Emoto's fascinating and inspiring research strongly suggests that torsion or subspace energy generated by human language can alter the structure of water. When the water in question surrounds the

DNA molecule—and all DNA in the body is surrounded by water—it stands to reason that such restructured water would have the vibratory power to transfer a new molecular configuration to the genome via transposons. In the case of tetrahedral water molecules (whose tripartite form recalls the trinity), this highly structured and energized liquid compound would penetrate the membrane of cells, harmonically signaling potential DNA to advance to the next stage of lightbody creation.

Potential DNA, in turn, projects a new unified frequency, one attuned to the universal background field of hyperdimensional torsion radiation emanating from Galactic Center, around the body. The ener-genetic precondition for lightbody expression, the Unified Consciousness Field is a gestalt that gradually becomes capable of resonating throughout at Source's signature tone of Ge—at which point the individual consciousness grounding itself in a luminous physiology can be considered unified with Source, or enlightened.

In his profoundly poetic and deeply philosophical *Starseed—The Third Millennium: Living in the Posthistoric World*, Carey discourses at length on the "unified field of consciousness," writing that the

> psychological process leading to awakening, though it can be described in many ways, is fundamentally a process of identity shift from a ... sense of self, rooted in your ego and an exaggerated sense of your vulnerability, to a sense of identity rooted in the unified field of consciousness that lies behind and within all individuality.

This field of torsion or conscious energy, what Wilcock calls the "consciousness field," is the true background of existence, giving rise to perceived multiplicity within the "illusion" of our world out of a fundamental unity. Continues Carey, "As you awaken into this awareness, you know yourself simultaneously as *one among many* and as *One at the source of many.*"

Amazingly, Leigh and I experienced Elucidation "bringing online" our ener-genetic "program" for the Unified Consciousness Field, which we so named at the time, long before either of us got around to reading Carey's books!

A first clarification regarding the Unified Conscious Field—at least as it is experienced, often quite powerfully, as a result of Elucidation—is that this field, in itself, does not equal our ascensional vehicle: the lightbody. Rather, the Unified Consciousness Field establishes and helps maintain a heightened unity consciousness by and through which our new physiology can come into being, supported at this time by heightened cosmic energy. Just as with healing, we cannot "enlighten" anyone else, but only offer facilitation so that the individual can move forward on his or her own path to God-realization.

The second clarification that needs to be made here is that while the Unified Consciousness Field is experienced as a gestalt, and often viewed by clairvoyants as pure white light (as explained below), this field—which as a micocrosmic representation of Galactic Center or the Creator, *is all things*—remains a composite of eight electromagnetic fields and chakras that are continuously in the process of balancing themselves.

In other words, and to expand on the above point, the Unified Consciousness Field is a result of a constantly running ener-genetic program—uploaded in this case through Elucidation—designed to keep the bioenergy centers (which in one sense still exist as individual centers) in a state of harmony. Only when the energy centers are balanced sufficiently can further evolutionary development, such as that initiated by Transcension Bioenergy Crystallization, occur.

I offer this somewhat complex understanding of the Unified Consciousness Field as an example of divine paradox—of the One simultaneously being the many, and vice versa. The bio-spiritual balance achieved and maintained via the Unified Consciousness Field can be understood even better, perhaps, using the analogy of the rainbow, which serves as a model of creation.

Each electromagnetic field and corresponding chakra (as well as each corresponding dimension) has a native harmonic frequency that manifests as a true color of the rainbow. Many esoteric traditions refer to these colors as *rays*. Starting with the first field/chakra/dimension, these colors/rays are always the classic ROYGBIV of high school science: red, orange, yellow, green, blue, indigo, and violet.

Unity consciousness is a very real ener-genetic phenomenon, encouraging conscious personal mastery and a more self-realized way of being, experienced by a relatively evolved individual.

Importantly, the eighth "color" in this model is the amalgam of all true colors—white—and corresponds, naturally, to the eighth or lightbody field/chakra (responsible for infusing the Unified Consciousness Field with Source energy or the tone of Ge), which links directly to the Creator/Galactic Center because, in a nonlocal sense, it *is* the Creator/Galactic Center. Unity consciousness here can be grasped as a very real ener-genetic phenomenon, encouraging conscious personal mastery and a more self-realized way of being, experienced by a relatively evolved individual.

As in Potentiation, Articulation and Elucidation, Transcension employs the five vowels to stimulate a self-healing potential in human genetics by way of the five corresponding nucleotide bases of DNA and RNA. The principal difference (other than the time it takes to perform Transcension, forty-five minutes as opposed to thirty minutes for the other sessions) is that the sound and light codes used in Transcension are more crystallized (ordered and harmonious) than in the previous activations.

This is exactly as it should be, since Regenetics is a technique for progressively stimulating consciousness and healing in a safe, integrated and integratable manner (Figure 5). By focusing on the spiritual body, which both experientially and according to *The Law of One* acts as a "consciousness shuttle" between our transdimensional Source and multidimensional incarnation, Transcension goes quite beyond even Elucidation to encourage a thoroughgoing transmutation of the emotional, mental and physical bodies.

The effect of this final DNA activation centered on the spiritual body is to draw in torsion energy directly from Source to transform the individual's electromagnetic fields/chakras, that are maintained in balance through the Unified Consciousness Field, starting with the first and progressing through the eighth—crystallizing (activating to a higher order of consciousness) the fields/chakras in order to assist the individual in embodying a truly advanced degree of conscious personal mastery.

In considering the role Source, Galactic Center or the Creator plays in this exponential process of personal evolution, I wish to propose another paradox: the "tone" of Ge really is not a frequency as we understand the term, but rather absolute *Silence*, the omnipotent, creational *Stillness* of unconditional love that differentiates into torsion waves of sound and then light during the threefold process of universal manifestation.

Stated a bit differently, Silent Stillness, pure creative potential, is the generative power behind the "primal sound" or Word that is responsible for directing construction of our light-based reality. When I consider this mind-bending subject, I am reminded of the many musicians who insist that contrary to popular misconception, it is not the notes themselves, but the intervals between the notes that make the music come alive.

The void state, Silent Stillness, what is called the Logos in *The Law of One*, is the primal formative matrix and "proportionately our most fundamental reality," Iona Miller and Richard Miller remind us in "From Helix to Hologram." "In essence, we emerge from pre-geometrically structured nothingness. DNA is the projector" of our reality from the embryonic stage forward.

Carey writes from a similar assumption regarding the chief role of sound—produced quite literally *vocally* by the Creator—in creation, including creation of "hologramatic biology." Employing his typical literary style, Carey explains that divine consciousness "began with the smallest possible particle, a tiny flash of autonomous *tonal insight* where several infinitesimal sound waves form[ed] a crossroads in the subatomic structure of what later became the nucleotides of DNA" (my emphasis).

In another article published in *DNA Monthly* entitled "The Universe Is Obsolete: A Gallery of Multiverse Theories," Iona Miller and Richard Miller theorize on the generative role of Silent Stillness from a bird's-eye, cosmological perspective. "Sound waves originated in the first instant of the universe's life," they propose.

"Vacuum-spawned particles flickering into existence from the Void were energized ... to remain in the real world. This sudden influx of countless particles from the vacuum was like throwing a stone into the dense particle pond of the early universe. Pressure waves rippling through the gas were nothing more than sound waves. The entire universe rang like a bell." The authors go on to postulate that the particle fog then "cleared and the universe became transparent. There was no longer enough pressure to support the sound waves. But now photons traveled freely through space ('Let there be Light')."

Another article reprinted in *DNA Monthly* further addresses the relationship between sound and light. In "Music to the Ears: The Infrared Frequencies of DNA Bases," award-winning composer Susan Alexjander relates how, with the assistance of renowned biologist David Deamer, she was able to measure the infrared energies of DNA nucleotide bases, translate them into sound, and thereby create an unprecedented type of DNA music. Alexjander explains,

> An octave in light is the same ratio as an octave in sound—2:1. A perfect fifth, or a relationship of 3:2, is the same proportion in light as in sound and can be interpolated to the world of geometry, architecture, movements of the planets, and so forth, so long as there is a periodic or regular vibration. By discovering patterns of ratios in light, we are simply translating into a sound

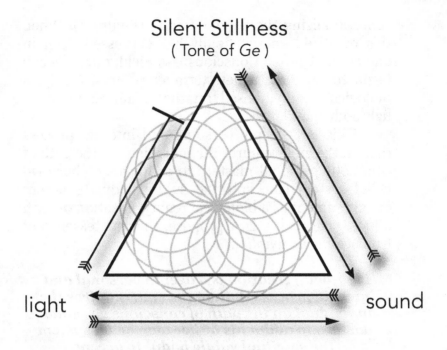

Figure 9: The Sacred Trinity
The image above illustrates the threefold process of universal manifestation in which Silent Stillness (the "tone" of *Ge*) gives rise to sound, which then translates into light (the holographic multiverse). Note that to return "home" to Source, we must retrace our steps from light to sound. Attempting to achieve bio-spiritual enlightenment without embracing the Audible Life Stream is a short-cut that leads to a dead-end.

medium to "hear" information and assess interrelationships. It could also be argued that both light and sound refer back to a common archetype which, as yet, is unknown to us, not unlike cousins who relate back to a common relative.

This archetypal "common relative" of light and sound, I contend, is Silent Stillness.

The divine triune structure of Silent Stillness giving rise to sound (the Word or Holy Spirit) which then becomes light (the sun or Son) is one way of conceptualizing the Sacred Trinity (Figure 9). When this trinitized energy finally has expressed itself in and as the Unified Consciousness Field, our cells can begin to replicate a new tetrahedral crystallinity, a biological trinity, based on the Adam Kadmon or lightbody template.

This is a deeply personal and internal process that, in the end, requires the individual on the path of conscious personal mastery to adapt his or her own belief system (the spiritual subtle body) to accept enlightenment as a biological reality. Creation of such a biological reality is, I propose, the quintessence of true healing or *wholing*.

> *Lightbody creation is a deeply personal and internal process that, in the end, requires the individual on the path of conscious personal mastery to adapt his or her own belief system (the spiritual subtle body) to accept enlightenment as a biological reality. Creation of such a biological reality is, I propose, the quintessence of true healing or* wholing.

The creational trinity composed of nothingness, sound and light is not unique to the Western worldview; it also is foregrounded in many Eastern philosophies. In Taoism, to cite one example, Silent Stillness is referred to as the *Tao*. The Tao gives rise to what Lao Tzu, author of the *Tao Te Ching*, calls the "ten thousand things" (the light-based holographic multiverse) by way of the breath (sound). From this perspective, Lao Tzu's references to "Immortals" are likely not metaphors, but indicate those "sages" who have arrived at bio-spiritual transcendence, or "holy flesh," by activating their lightbodies.

Polich, calling the divine lightbody blueprint the "god-seed," offers a helpful analogy for how the Adam Kadmon unfolds both individually and collectively: the "seed of any flower has within it a hidden code contained in the plant's DNA that becomes activated under certain conditions ... Similarly, the light codes within the human form may be viewed as spiritual DNA ... the inner forces that trigger the process of enlightenment. The ultimate maturation of the god-seed ... is divinity, an example of which is Christ consciousness."

On a species level, embodying Christ or unity consciousness implies being capable of naturally utilizing hypercommunication. In the animal kingdom, hypercommunication has been employed successfully in a variety of species for millions of years. Fosar and Bludorf point to the minutely orchestrated flow of life in insect colonies as an example.

When a queen ant is taken from her colony, construction normally continues. If the queen dies, however, all work stops. Apparently, the queen sends the "construction blueprint" even from great distances via the collective consciousness/DNA of her colony. She can be as far-off as she likes, so long as she remains alive.

Fosar and Bludorf compellingly argue that during prehistory humanity was, like most animal species, closely connected to group consciousness. This period equates to the "second density," vibrationally designed for animalistic minds and bodies to experience a type of collective awareness, in *The Law of One*.

To evolve and experience individuality, and ultimately individuation, humans chose to abandon hypercommunication almost completely. Thus we entered the "third density," at the end of which we now find ourselves preparing to enter the "fourth density," where our focus as a species will transcend self-centered behavior rooted in survival fears and become both more unified and heart-based.

Significantly, my research indicates that the third-density cycle began with creation of the Fragmentary Body as a means of differentiating self from other and embarking on a path of individualization leading to individuation. My understanding of this phenomenon is that our ancestors' "decision" to give up hypercommunication helped create "the Fall," which rather than being an irreversible cataclysm, was actually a crucial step in the evolutionary process leading to greater and greater levels of embodied light or consciousness.[20]

Now that we are relatively grounded in our individual consciousness, thanks to this crucible of separation and fragmentation we blithely call reality, we are in a position to create a new network of

---

[20] In her book *Catastrophobia*, Barbara Hand Clow makes a case for a major global catastrophe (apparently caused, at least substantially, by a supernova eruption in the Vela system) about 11,500 years ago. Clow argues that the trauma was so great and widespread most humans shut down their intuitive side. While there is evidence for such a catastrophe, this theoretical event did not create the Fragmentary Body. Although Earth's third-density cycle dates back approximately 75,000 years (comprising three cycles of 25,000 years each) according to *The Law of One*, the precise origins of the Fragmentary Body are difficult to pinpoint, as it appears to be a key aspect of the archetypal system of our Logos (Galactic Center). As such, the Fragmentary Body is linked closely to the "veiling" (forgetting) we experience as we incarnate in the third density—on whatever planet in our galaxy, at whatever time. As we evolve out of the animalistic consciousness of the second density, the Fragmentary Body is a rift that automatically opens to create the necessary "spacing" for growth of individual consciousness leading to conscious reunification with the Creator. This rift can be widened, just as it can be lessened, and the type of catastrophe Clow explores can be seen as an exacerbation of this rift—but not its cause.

collective consciousness: one in which we learn and communicate through our DNA—a conclusion with many similarities to Sheldrake's concept of Morphic Resonance. Just as with the Internet, Fosar and Bludorf demonstrate, *one can "upload" data into the biological Internet that is DNA, "download" data from it, and even "email" other participants.*

It is my personal belief, one I share with Leigh and perhaps millions of others, that the time is just on the horizon for the maturation of the god-seed and global lightbody activation leading to a hypercommunication revolution with the ability to unite all of humanity as a differentiated, enlightened consciousness. "A moment is coming after which nothing will ever again be thought of as it was before," writes Carey. Quickly approaching is "a metahistorical moment, an event simultaneously alpha and omega to [our] species and all [our] species has ever known."

"The age that has been written about, whispered about, and spoken about is upon you," observes Marciniak. "It is the age when humanity physically mutates ... and literally turns into something ... it was not a short time before ... multidimensional beings." For the ancient Incas, our generation, which is fortunate enough to experience the "end of history" with the Galactic Alignment of 2012, is poised to become the true *chakarunas*, "the bridge people" tasked with creating Heaven on Earth.

14
The Biology of Enlightenment

The evolution from human to divine consciousness involves healing the perception of duality and its legacy of karma and disease at the cellular and atomic levels. There is perhaps no illness that cannot be healed through the proper exercise of intention. Many of the thousands of documented so-called miracle healings powerfully demonstrate the impact of consciousness on physical as well as emotional, mental and spiritual wellbeing.

Mind-body medicine, which is statistically valid enough to be taught in today's medical schools, offers additional proof of our ability to heal ourselves. Bruce Lipton's research, as detailed in *The Biology of Belief*, further indicates that people can modify their DNA and overcome life-threatening illnesses simply by changing their consciousness in general, and their beliefs about health and wellness specifically. Gregg Braden's *The Spontaneous Healing of Belief: Shattering the Paradigm of False Limits* continues this life-changing line of thinking.

Deepak Chopra has remarked that "the similarity between a thought and a photon is very deep." A photon is a particle or quantum of light or other electromagnetic radiation. Dr. Chopra is implying a close connection between thought and light.

To reiterate, in the Regenetics model, *thought (intention) is considered a form or function of torsion energy manifesting as hyperdimensional light*. Mind is "the illuminating energy which 'Lights the way' of an idea or form to be transmitted and received," wrote Alice Bailey. "Upon a beam of light can the energy of the mind materialize." Following this line of reasoning, we can imagine ourselves not only as frozen light (recalling Richard Gerber), but also as "frozen thought."

Looking at the human body as a congealed thought, which at first may strike the reader as strange, is in the final analysis deeply empowering. Repeatedly, quantum physicists have demonstrated that a scientist always and inevitably affects the outcome of an experiment simply by observing it, a realization now universally accepted in the scientific community as the Heisenberg Uncertainty Principle.

Even more amazing is the paradigm-altering discovery that gave rise to the aforementioned particle-wave duality: the probability that the physicist actually *creates* the quantum particles he or she observes, since in unobserved states these particles appear to exist only as waves.

A fundamental and revolutionary truth emerges from this information: *consciousness creates*. As human beings imbued with free will, we can use the power of our consciousness to re-create our reality: including but not limited to a body, mind and spirit free of disease.

I emphasize "re-create" because, clearly, we already inhabit one creation. The world as we know it is based on the principle of duality. Another way of stating this is that a dualized or divided consciousness, one that already saw itself as separate from other consciousnesses, including unity or God consciousness, gave birth to the universe as humans typically experience it: a battleground between good and evil, light and dark, right and wrong, "us" and "them."

But duality is not merely a philosophy; it is a physical state of being as well. The very atoms that make up our cells are founded on positive and negative charges whose opposition sustains a certain life-form. Years ago, Lipton coined the phrase "biology of consciousness" to summarize the transformational idea that living organisms, including humans, rather than being empirical givens, are actually malleable *thought-forms*. In other words, adopting a quantum perspective, we are basically waves that only cohere as particles through an act of consciousness. By changing our consciousness, we change our physical form and functioning.

> *Lipton coined the phrase "biology of consciousness" to summarize the transformational idea that living organisms, including humans, rather than being empirical givens, are actually malleable* thought-forms. *In other words, adopting a quantum perspective, we are basically waves that only cohere as particles through an act of consciousness. By changing our consciousness, we change our physical form and functioning.*

Healing, as previously pointed out, means to make whole. Healing ultimately results in unification or enlightenment and implies atonement, which in this context should be read as "at-one-ment."[21] In a world where thought creates and biology is a product of consciousness, not the other way around, the mind has the power to forge a new biology, one no longer

---

[21] According to Horowitz, *atonement* "has at least three important meanings: 1) 'at-one-ment,' or becoming *one* with the Creator; 2) 'a-tone-meant,' or an *intended* sound, electromagnetic frequency, or mathematical vibration [and] 3) 'a-tone-meant,' or the *meaning* of a tone reflecting the meaning of life as a creative concert of life-sustaining ... sound, electromagnetic frequencies, and bioacoustic resonances interacting with matter."

based on the illusion of duality, but on principles of unity and harmony that are characteristics of universal truth.

In *Return of the Bird Tribes*, where a central theme is the reunion of the human body with its animating consciousness or angelic aspect in the pivotal years we currently are experiencing, Ken Carey neatly summarizes how we must proceed, individually as well as collectively: "In the order of healing, it is human consciousness that first must change." *Our challenge, which is also a tremendous opportunity, is to open up to a literally life-changing way of thinking ourselves into existence.*

Enlightenment, as elaborated in previous chapters, is the process of raising consciousness and letting the light of higher awareness in to the point that we *embody* it. True enlightenment follows a path of conscious personal mastery that results in transformation and, by definition, involves creation of a lightbody. The lightbody is a trinitized (balanced and harmonious) physical vehicle that has resolved the fragmented perception of duality, karma and disease at the cellular and atomic levels.

We can conceptualize the current evolutionary Shift in our species' DNA as a change in operating systems from a binary to a "trinary" code based on the ener-genetics of the threefold tetrahedron shape. We might go so far as to say humans are evolving out of biology into "triology." In this light, it is most interesting to recall that some members of the alternative science community have alluded to suppressed research on a third DNA strand reportedly activating in many humans (Figure 7).

An illuminating way of visualizing how metamorphosis into a unified, light-based physiology occurs is to look at a quantum particle known as positronium. Positronium is composed of an electron, which has a negative charge, and a positron, which has a positive charge.

We can conceptualize the current evolutionary Shift in our species' DNA as a change in operating systems from a binary to a "trinary" code based on the ener-genetics of the threefold tetrahedron shape. We might go so far as to say humans are evolving out of biology into "triology."

Positronium is a perfect example of duality. It also provides a wonderful illustration of how the lightbody is created. Since electrons and positrons are antiparticle opposites, after combining to form positronium, they immediately cancel out each other and decay into two particles or quanta of light (photons). A third stable and unified "element," which is neither positive nor negative, thus is created from a preexisting dualism.

Barbara Hand Clow writes that this process of combination and decay in the positronium atom, mirrored in lightbody activation, "resolves inherent duality into light ... [As] the electron is the basic unit of activation—life—it triggers the transmutation of the positron—karma." Contrary to popular misconception, karma has nothing to do with punishment and reward. It exists as part of our holographic universe's binary or dualistic operating system only to teach us responsibility for our creations—and *all* things we experience are our creations.

When our creations are unconscious and thus out of tune with Source, they often manifest in the disharmony known as disease. This can occur not only in individuals, but in entire civilizations. In both cases, disease, which typically is considered a crisis, simultaneously serves as a powerful stimulus for transformation and transcendence. Today we see this process occurring all around us.

As we raise our consciousness and assist the cosmos in activating our higher-dimensional lightbody, we realize we are our own creators made, or making ourselves, in the image and similitude of the one Creator. Indeed, since in a hologram the part contains the whole, we *are* the one Creator.

By learning this truly transformative lesson, we return to unity consciousness while, to a considerable degree, mastering physicality. Otherwise stated, we achieve an expression of God-realization as the light of higher consciousness descends into a divine lightbody healed of the grosser distortions of duality and freed from the instructional cycle of karma.

This process of resolving duality and its legacy requires sealing—and eventually healing—the Fragmentary Body, a process which can be initiated with a specific meta-genetic activation and bioenergy recalibration like that accomplished through Potentiation Electromagnetic Repatterning.

If consciousness creates, and reality including biology is a thought-form, and in the beginning was indeed the Word, then it is critical we realize that a divided consciousness (being an aspect of our own consciousness) employing some type of divisive *sound* combined with a separatist *intention* gave birth to our ostensibly dualistic universe.

The notion that we are children of a lesser god (or lesser aspect of ourselves) goes back at least as far as the Gnostics, whose humane yet cosmic philosophy profoundly inspired Jesus. To summarize, the Gnostics believed that the universe and its inhabitants are imperfect (i.e., unhealed) manifestations arising from a false sense of separation from Source—the Original Wound—and that the ultimate goal of life on Earth is to return to a state of enlightenment or wholeness.

Citing a line from William Blake, "God is man and exists in us and we in him," Braden in *The God Code* exhibits a philosophy very much in line with Gnostic thought when he reiterates, "God exists as humankind so that humankind may make the choices each day that bring us closer to the perfection of our Creator."

Elsewhere, Braden's Gnostic orientation appears even more pronounced. "The Book of Genesis provides some of the most powerful clues to understanding our role in the destiny of our species," he writes. "In its original language of Hebrew, the text reveals that during the act of creation, God stopped the process *before* it was complete (Genesis 17:1). The English translation of this telling event reads: 'I am God the Creator who said enough, now walk before me and *become* perfect.'"

Clow displays a similar "Gnostic" cosmology when writing about what has been called the Fall: "You walked out of the Garden of Eden ... and ... split your world by viewing it through your eyes and brain instead of feeling it in your heart. The greenness of your world became separate from you, and time began." Clow also associates the Fall with speech. After humanity attempted to name the Creator, she writes, "then everything had to have a name; *language began as an identification process instead of using sound as a resonation tool ... for perceptual fusion*" such as that characteristic of a system of hypercommunication.

From a somewhat more scholarly angle, Charles Eisenstein argues essentially the same point. "The ascent of humanity is in part a descent into a language of conventional symbols, representations of reality instead of the integrated vocal dimension," he explains. "This gradual distancing, in which and through which language assumed a mediatory function, paralleled, contributed to, and resulted from the generalized separation of man and nature. It is the discrete and separate self that desires to name the things of nature, or that could even conceive of doing so." Specifying that to name "is to dominate, to categorize, to subjugate and, quite literally, to objectify," Eisenstein concludes, "No wonder in Genesis, Adam's first act in confirmation of his God-given dominion over the animals is to name them."

Similarly, in *Healing Sounds* Jonathan Goldman relates,

> There are legends that before there was a spoken language of words, there was a harmonic language. This language allowed humankind to communicate with all the creations of nature [utilizing] the concept of information ... encoded on ... pure tone frequencies ... This may be one of the ways ... dolphins communicate [by] transmitting three-dimensional holographic thought-forms of sound. Eventually this harmonic language separated [when] consonants combined with the tones to create words.

These legends, like the basic cosmology behind nearly all religions and mythologies, prioritize the creational energy of sound combined with intention. Together, these form a divine language of creation called by alchemists the Green Language and known in the Koran as the Language of the Birds. To paraphrase another of Goldman's equations, vocalization + visualization = manifestation, we can affirm that

SOUND + INTENTION = CREATION.

From our previous discussion of the literal alphabet that is DNA, it should be obvious that science is beginning to confirm this belief in the life-giving primordial Word (pure torsion energy unfolding to create a generative bioacoustic and electromagnetic interface consciously imbued with meaning) that provides the foundation for the overwhelming majority of the world's spiritual belief systems.

When discussing Regenetics, Leigh and I regularly refer to sound and intention. From the time we began developing this work, we intuitively felt that of these two energies, sound is primary. Only later did we come to understand that sound derives directly from the unified torsion energy of unconditional love, the Silent Stillness sometimes reached briefly in meditation.

The idea that light stems from sound may strike some readers as contrary to what is observed in nature, where sound operates at a measurably lower vibratory rate than light. Here, however, we are discussing sound and light *in higher-dimensional states* where other laws of physics apply. Arguably, *opposite* laws—rather, the same laws running in reverse order—apply, given the cosmic mirroring inherent in the concept of "As above, so below."

Richard Miller and Iona Miller are two of a growing number of researchers who believe that the creational energy emanating from Galactic Center (Ge) begins as sound and manifests as light that creates the holographic illusion of reality—a translation process detailed in Chapter Six that is replicated precisely in the human body at the level of chromosomes. The thirteenth-century Persian saint Shamas-i-Tabriz shared a similar belief, poetically proposing that

The Universe was manifested out of

The Divine Sound;
From It came into being the Light.

Perfectly consistent with this line of reasoning, the residual energy of the so-called Big Bang, at first dismissed as "background noise," was picked up by *radio* telescopes. "The cosmos is filled with sounds and rhythms, from pulsars and quasars, from supernovas, from so-called 'red giants' and 'white dwarves,' from fleeting and colliding star systems, from our own sun," writes Joachim-Ernst Berendt.[22]

The influential Indian philosophy of *Nada Brahma* literally translates as "the world is sound." In the Upanishads, according to Berendt, the "principle of Brahma, the so-called *Brahman*, is the prime power of the cosmos." It is written that "Brahman is the absolute./Everything that exists is Brahman or the Sacred Word." Here, the striking and undeniable similarities between Brahman, the Tao and the tone of Ge should be obvious. The thoroughly logical point I am making is that if everything is

---

[22] The cosmological model of creation I am proposing in which sound gives rise to light not only is consistent with the majority of world religions and mythologies and the genetic sound-light translation mechanism. This model also may help bridge the gap currently dividing those physicists invested in dark matter and those who follow the Electric Universe theory. The latter contend that the existence of plasma as an operative principle in cosmology does away with the necessity for looking to other dimensions for missing matter. If, however, we consider plasma "liquefied light," the form light takes during manifestation, both theories may be essentially correct. Since matter and energy are interchangeable, we can imagine a continuous movement of torsion energy in the form of hyperdimensional light into our own dimension, where it manifests as plasma before becoming "physical."

somehow sound, how can light *not* derive from sound?

Some scholars, such as Henry and the Weinholds (as noted), go so far as to argue that the famous quest for the Holy Grail is actually to attune our DNA so that it resonates with Galactic Center's tone of Ge. This divine "frequency" capable of generating a "love song in the blood," a "Song Grail" or *San Graal* ("Holy Blood"), can attune the core of our being, our DNA, to the galactic Core's transformational "vibration" of unconditional love.

If everything is somehow sound, how can light not derive from sound?

To transcend our multidimensional existence based in light and reunite with our transdimensional Source in the Silent Stillness that gives rise to sound, then, appears to be the final goal of the true spiritual seeker or adept. That we walk this evolutionary path by incarnating in physical vehicles known as lightbodies, their dimensional differences notwithstanding, poetically confirms that we travel the path of light until we encounter—and become—the Word.

"The point comes in the life of every sincere spiritual seeker," writes Dennis Holtje in a fascinating book entitled *From Light to Sound: The Spiritual Progression*, "when information, knowledge, and finespun theory no longer satisfy our innate quest for spiritual liberation ... Should we desire a more fulfilling ... relationship with soul and the God within, we must surpass the reach of this light energy and experience the spiritual energy which parents it—the Sound Current."

In other words, light is about information whereas sound is about *transformation*. We can spend lifetimes following the light trying one spiritual and/or intellectual technique after another, but until we embrace what Holtje elsewhere calls the Audible Life Stream, we never will find our way home. In a similar vein, *The Law of One* emphatically states that in our current third-dimensional experience, it is impossible truly to "know" anything. Rather, as a prerequisite for moving beyond this reality, we are here to develop compassion and love—for self as well as perceived others—through an ongoing exercise of faith based in *intuition* or *feeling*.

Light is about information whereas sound is about **transformation.** *We can spend lifetimes following the light trying one spiritual and/or intellectual technique after another, but until we embrace the Audible Life Stream, we never will find our way home.*

The point I am stressing is that although it can be discussed, enlightenment is not a cerebral process and cannot be reasoned out; it must be *experienced* at the creational level of sound. "The stunning simplicity of the Sound energy confounds the mind," explains Holtje. "We are conditioned to use the mind to solve all of life's dilemmas, unaware that the latent energy of Sound, once released, provides the permanent solution of awakened spiritual living." With sound, writes Marciniak on a related note, "it is quite easy to bypass the logical mind." Moreover, the evolution of "DNA expresses itself beyond logic through sound."

While affirming that "biology must merge with love" if humans are to evolve what we might call our "triology of consciousness," Clow simultaneously implies that the infinite creational energy of Source is ultimately unknowable. *The Law of One* also states that our evolutionary journey away from and back to the Creator "begins and ends in mystery."

At the same time, we have the very "Law of One" notion of god-seed and divine self-realization that sees the current stage of human evolution as a process of "meeting our Maker" while still very much alive. From this perspective, the fundamental reality we now are being called to embody is that we are one with the Maker. Indeed, we *are* our Maker.

While making space for the divine mystery of being to remain mysterious, we can see clearly that a holistic perception results from removing the distorted lens of duality, which is accomplished through opening our hearts to unconditional love and allowing ourselves to evolve into unity consciousness and its corresponding physiology of light—an epistemological and ontological enlightenment that at some point necessarily involves healing the *perception* of duality.

In turn, healing the perception of duality allows us to resolve duality in our spiritual, emotional, mental and physical bodies by incarnating a higher avatar of our soul (whose "sole" principle is unity) in the form of the lightbody.[23]

---

[23] The student of *The Law of One* will note that I have simplified somewhat the notion of lightbody in this book. In truth, the so-called physical vehicle for consciousness appropriate to each dimension or density above our own is a form of lightbody. This means that we inhabit, in reality, a series of increasingly "conscious" or light-filled lightbodies on our evolutionary journey back to complete, undifferentiated Oneness with the Creator. For the most part, when the lightbody is discussed, I am referring to the fourth-density form that will be activating for the majority of those "harvestable" sometime around 2012.

15
"So Below, As Above"

Given that the very atoms that compose matter as we know it are dualized into positive and negative, it should come as no surprise that duality also expresses itself in the body's electromagnetic fields. This brings us back to the critically important but largely ignored concept of the Fragmentary Body occasionally encountered in theosophical teachings and introduced in Chapter Seven.

Through muscle testing on hundreds of clients, as related in Part I, Leigh and I found that the second electromagnetic field (which corresponds to the second or sex chakra) exists in most people as a bioenergy or consciousness vacuum that to a significant degree separates spirit and matter.

The Fragmentary Body does this by draining one's kundalini (the individual's own life-wave of torsion energy) as it seeks to rise into the higher electromagnetic fields and chakras while impeding the free flow of torsion energy or universal creative consciousness emanating from the "Core" of our being in Galactic Center that seeks to infuse the lower fields and chakras. The Fragmentary Body represents a major impediment to the sacred marriage between the lower self and Higher Self, or the embodiment of the light of higher consciousness that occurs during genuine enlightenment.

The Fragmentary Body constitutes a torsion "rift" in the human bioenergy fields that corresponds to the Great Rift dividing the consciousness that—in order to promote our evolution—created the system of duality, in some traditions visibly manifest in the night sky as the band of the Milky Way. Such mirroring is entirely in keeping with "As above, so below," since this holographic principle that conflates the part with the whole requires that our consciousness-based physiology be consistent with the cosmological consciousness that created and sustains us.

The Fragmentary Body is associated not only with sexuality, but also with the mouth and specifically the tongue: our speech organs. This connection further supports my contention that *our dualistic reality literally was spoken or sung into being by a higher aspect of ourselves*. The primacy of speech in creation is perhaps why the Egyptian goddess Nut, a feminine icon for Galactic Center, is pictured in hieroglyphics as a great milk cow with full *udders*. Today many scientists and theologians agree that if we indeed were created, we were, to borrow a William Henry pun, *uttered* into existence.

During the time leading up to the Tower of Babel, which I see as a metaphor for the perceptually fused or unified state of consciousness (hypercommunication) that preceded humanity's experimental and educational Fall from grace that created optimal conditions for individuation, the Bible records that "the whole Earth was of one language, and of one speech."

The stories of the Garden of Eden and Tower of Babel are of special interest because they associate human biology with language. While the language of the Garden can be read as divisive, resulting in the Fall with its fallen biology, that of the Tower initially appears both unified and unifying, leading upward (not unlike a vertical DNA helix, the Sacred Spiral) to a heavenly marriage between human and divine.

The divine language of the Tower, spoken by Jesus and called the Language of the Birds, the Green Language and also the *lingua adamica*, is composed exclusively of vowels.[24] This was a major reason vowels were extracted from written Hebrew: they were revered as sacrosanct, God's language of creation. The "vowels were 'extras' in Hebrew," writes William Gray in *The Talking Tree*. "The vowels were originally very special sonics indeed, being mostly used for God-names and other sacred purposes. Consonants gave words their bodies, but vowels alone put soul into them."

In *The Law of One*, any sound healing catalyst that properly employs vowels is described as a "scalpel" compared to the "blunt instrument" of other techniques for encouraging conscious evolution. Over the years, many clients of the Regenetics Method have referred to this form of vowel-based DNA activation as "instantaneous karma," indicating a way of very, very rapidly clearing and integrating (i.e., learning and healing from) possibly lifetimes of unprocessed experiences capable of expressing themselves in one or all of the physical, mental, emotional and spiritual bodies.

Compared to ritualistic forms of experiential catalyst such as prayers, affirmations, mantras, reiki and other predominantly light-based approaches that by repetition may have established a morphogenetic

---

[24] Medieval alchemists used the phrase "Green Language" probably owing to the critical alteration of hydrogen bonds (which are green on the electromagnetic spectrum) in biological water molecules that appears to occur during lightbody creation. Not surprisingly, many figures associated with the lightbody, such as Osiris and Quetzalcoatl, typically are depicted as green. Similarly, Lao Tzu's Immortals achieve bio-spiritual enlightenment by creating a "jade body." As this book refers primarily to our fourth-dimensional lightbody, it is fascinating to remark that the true color or ray for the fourth density responsible for engendering this particular lightbody, according to *The Law of One*, is also green.

field that can be tapped into for certain healing purposes, *The Law of One* states that vowels "have power before time and space and represent configurations of light which built all that there is."

In other words, certain vowel sequences, apparently existing in a transdimensional state within Galactic Center or the Logos *outside* the multidimensional Akashic record, stem directly—with no intermediary or dimensional "translation"—from the Mind of the Creator steeped in Silent Stillness. As such, these vowel sequences, forming the Language of the Birds, have power above and beyond most energetic modalities to reset, on a permanent basis, an individual's bioenergy matrix—inviting a quantum influx of the Creator's consciousness, which is always already whole and intact, and tremendously encouraging healing or wholing on all levels.

At this stage, it should come as little surprise that the higher-dimensional consciousness behind *The Law of One* calls itself Ra. To the ancient Egyptians, the bird-headed Ra was the sun god believed to have created the world through speech.

"In the case of ... Hebrew," explains Ra, "that entity known as Yahweh aided this knowledge [of sacred creational vowels from outside dimensional expression] *through impression upon the material of genetic coding which became language*" (my emphasis). Nearly two decades before the fact, this statement from 1982 uncannily anticipates the findings in the late 1990s of wave-genetics—a revolutionary meta-genetic science (that is, above and beyond both genetics and epigenetics) which, to reiterate, clearly establishes that 1) human DNA is inherently "linguistic" in nature; and 2) human language can be adapted to activate and heal DNA and, thus, the entire organism.

It is absolutely astonishing that the Regenetics Method came to be, complete with specific vowel sequences that often stimulate miraculous results in clients, five years *before* Leigh and I began reading *The Law of One*. I also wish to note that in this book I barely have scratched the surface of the large number of extraordinary parallels between *The Law of One* and Regenetics, both of which recognize eight bioenergy centers, in the macrocosm as well as in the microcosm of the potentiated human being.

It appears that vowels indeed are sacred because, as represented by the five vowels in English, they correspond to the five nucleotide bases of DNA and RNA used to create biological organisms: adenine, cytosine, guanine, thymine, and uracil. These five nucleobases, represented as A, C, G, T and U, literally form our genetic alphabet.

Mastery of the Language of the Birds effectively turns one into a master geneticist not only capable of creating life-forms via speech, but of promoting evolution of these life-forms into light-forms *structured on the tripartite tetrahedron shape that is the basic building block of the* merkabah.

Mastery of the Language of the Birds effectively turns one into a master geneticist not only capable of creating life-forms via speech, but also of promoting evolution of these life-forms into *light-forms* structured on the tripartite tetrahedron shape that is the basic building block of the merkabah. This is consistent with Ken Carey's assertion that once you are "awakened," "you know your human circuitry for what it is, a system of empathy, representation, and creation, designed to *regulate and evoke biology*" (my emphasis).

The Fragmentary Body is electromagnetic evidence of "mispronunciation" of the Language of the Birds, apparently through insertion of consonants that disrupt the unified flow of vowels and create dualistic syllables. The story of the Garden of Eden appears to record this historical speech event that gave rise to the illusion of duality in the form of the Fragmentary Body—which energetically corresponds not only to the mouth and tongue, but also (appropriately given the procreative "doom" that results from Adam and Eve's fateful discussion of knowledge) the genitals.

The Fragmentary Body is electromagnetic evidence of "mispronunciation" of the Language of the Birds, apparently through insertion of consonants that break up the unified flow of vowels and create dualistic syllables.

With many parallels to Eckhart Tolle's concept of the "pain body" that keeps people from accessing the infinite "Power of Now," the Fragmentary Body operates much like a deep scratch in a record or, to use a Vedic term, a *samskara* that maintains one's consciousness locked in a limited (unenlightened) matrix of thought and belief banished from knowledge or gnosis of unity with the Garden or Ground of Being.

Toltec masters often refer to the Fragmentary Body as the Parasite—fittingly, since at the energetic level parasites (physical and aetheric) enter and establish themselves at the human organism's expense by way of the Fragmentary Body. This implies, of course, that sealing the Fragmentary Body is capable of expelling parasites from one's system—as commonly and dramatically is reported to occur over the course of Regenetics without recourse to therapeutic interventions such as drugs, parasite cleanses, or zappers.

In all cases, to be absolutely clear, until it is sealed ener-genetically, *the Fragmentary Body is an illusory—but no less powerful—dualistic principle of limitation that promotes disconnection from Source and, ultimately, death.*

While developing Potentiation Electromagnetic Repatterning, Leigh and I became increasingly interested in this problematic second bioenergy field. We also were intrigued by the fact that, using muscle testing, we consistently were finding nine electromagnetic fields corresponding to nine chakras. Our understanding from theosophy and Vedic teachings was that there were only seven fields and chakras. But thousands of kinesiological tests convinced us there are nine fields and nine chakras, in addition to a tenth "energy center" associated with Galactic Center and Silent Stillness.

We concluded that this tenth field (which generates the unifying tone of Ge) is where the Higher or God Self, also known as the soul or atmic permanent atom, resides. While the other four permanent atoms (physical, emotional, mental, and spiritual) are associated with particular lower electromagnetic fields as indicated in Appendix C, the atmic remains unified in the Master or Source Field.

The atmic permanent atom thus represents the divine or soul aspect of ourselves that never experiences fragmentation and loss of unity consciousness through the veiling associated with duality. In other words, the atmic permanent atom, or Adam, never was banished from the Garden. This is also the part of ourselves that now is seeking to transmute our consciousness and biology—via the *lingua adamica,* no less—so that we can evolve into what we always have been *in potentia*: the Adam Kadmon.

What happened next in our process of ener-genetic unfoldment was extremely exciting. After Leigh and I performed Potentiation on ourselves and began to experience what we came to understand as torsion energy from Galactic Center activating our DNA, we realized that two significant shifts, both of which we substantiated through kinesiology, were happening in our energy bodies.

First, the vibratory frequency of our electromagnetic fields began increasing noticeably. Second and more importantly, the fields themselves were undergoing a progressive recalibration from nine to eight in number.

In this process, the ninth field literally descended and sealed the second field or Fragmentary Body, allowing the spiritual energy of the Healing Sun to flow through our electromagnetic fields and chakras uninterrupted for the first time. It was at this point (around the five-month mark following Potentiation) that my health began to improve dramatically and Leigh's asthma and environmental allergies completely disappeared.

In retrospect, this energetic repatterning makes perfect sense. By bridging the rift in our bioenergy fields created by the Fragmentary Body, we were taking a first step toward healing the perception of duality at the level of our biology.

We also were recalibrating electromagnetically to eight energy centers, which represents a critical alchemical phase of lightbody activation having to do with creating an infinity circuit in the bioenergy blueprint. The number 8 always has been of utmost importance to masters of the Language of the Birds. Thoth (often associated with Ra) was the Lord of 8, the Buddha taught the Eightfold Path to enlightenment, and numerologically, the name Jeshua or Jehoshuah (Jesus) translates as 888.

In her discussion of duality and karma, Clow makes a highly intriguing observation. The centerpiece of her cosmology is the Photon Band, which, to reiterate, can be visualized as a hyperdimensional torsion lattice of light connecting Earth to Galactic Center via the sun that serves as a guiding data communication network for human and planetary evolution (Figure 6).

As we release karma and create our lightbody, Clow writes, our positrons and *"electrons collide, quanta of light are formed, and the Photon Band manifests!"* In other words, we are not merely passive receivers of evolutionary energy; we also are active creators of this very same energy.

We are not merely passive receivers of evolutionary energy; we also are active creators of this very same energy. Through consciously infusing our being with torsion waves of kundalini, we partner with the galactic evolutionary plan being directed from "above" by facilitating our personal enlightenment from "below."

Through consciously infusing our being with torsion waves of kundalini, we partner with the galactic evolutionary plan being directed from "above" by facilitating our personal enlightenment from "below." This happens as we commit to healing our own distortions with regard to duality and begin to operate out of unity consciousness and unconditional love, which gradually resolves conflict and disharmony in all levels of our being—from the genetic up—as we individually return to harmonic identification with Source.

By making the conscious commitment to find ways to attune ourselves to Galactic Center's torsion energy emissions, specifically the tone of Ge that stimulates our DNA to evolve our enlightened physiology from within, we personally perform the godlike task of ushering in the Age of Consciousness. As inherently spiritual beings with an innate impulse to return home, we are calling to us the very spiritual energies designed to take us there, proving that "As above, so below" can, and should, be read the other way as well: "So below, as above."

This truthful paradox dissolves at a stroke the false opposition between "nature" (above) and "nurture" (below), inviting us to appreciate the intricate evolutionary interplay between the macro- and microcosmic. Just as importantly, this paradox also invites us to surrender any residual belief not only in causality, but in our own powerlessness, and embrace a new way of thinking in which a single person in touch with the divinity within can alter the course of history by forging a new reality.

"As I do these things, so shall ye do them, and greater things," said Jesus, who also stated, "Nothing will be impossible for you." Acknowledging our inner divinity as a step on the path to embodying it is not to be confused with narcissism or individualism, since we must admit that everyone's divine birthright is the same limitless creational potential of unconditional love.

As mentioned near the beginning of this book in Chapter One, in *The Isaiah Effect* Braden claims on the basis of his study of one of the Dead Sea Scrolls that the Gnostics from the time of Christ employed a type of prayer called *active prayer* to alter quantum outcomes by changing the pray-er's picture of reality. Active prayer, as Braden describes it, is a type of focused intention that appreciates whatever is being requested as already having occurred.

This is not the place to go into a full explanation of active prayer. I merely wish to point out that active prayer employs five interactive modalities that, together, are capable of changing quantum outcomes. Braden denominates the first three of these modalities *thought*, *feeling*, and *emotion*. If this trinity is utilized harmoniously and combined with *peace* and *love*, then "our world mirrors the effect of our prayer."

Of particular relevance to the Regenetics Method is that these five intercessory modalities (thought, feeling, emotion, peace, and love) correspond kinesiologically to the five nucleotide bases of DNA and RNA—which, in turn, have been shown to align with the five vowels. This means that when we use language such as prayer or an energized narrative to change reality consciously, we do so through our bodies by activating our divine genetic endowment: our quantum biology.

By indications too numerous to detail and only suggestively touched on in these pages, we are, in fact, approaching a fundamental, quantum Shift in the very nature of reality, in which even as I write these words the physical laws of the universe (including those governing the expression of our physical vehicles) are transforming.

In the words of Terrence McKenna, in whose Timewave Zero theory, with no conscious knowledge of the Mayan calendar, the year 2012 was envisioned as the "final time," we are closing in on a "necessary singularity, a domain or an event in which the ordinary laws of physics do not function [and in which we exit] one set of laws that are conditioning existence ... into another radically different set of laws. The universe is seen as a series of compartmentalized eras or epochs whose laws are quite different from one another, with transitions from one epoch to another occurring with unexpected suddenness."

In a multidimensional reality composed of infinite parallel universes, any of which suddenly can land in our own like a ball bouncing on a roulette wheel, we can participate actively in changing the future by simply, following John English in *The Shift*, dreaming the one we want into being.

"Rather than *creating* our reality," Braden has suggested, "it may be more accurate to say that we create the conditions into which we *attract future outcomes*, already established, into the focus of the present." Marciniak conceptualizes this aspect of conscious personal mastery having to do with manifestation as "reality adjusting" and understands it as an inherently energetic endeavor: "Refocusing your attention to reinforce the outcome you desire will alter the frequency you transmit, inevitably opening the door to another probable outcome."

Conscious personal mastery is a deeply personal and internal process that always occurs in the present. In a world such as ours with foundations in free will, no technology, teacher, guru or savior, however advanced, can do for us what only a commitment to evolving and operating out of our own divine consciousness and physiology can achieve.

In the final analysis, the choice is ours in every now whether to give away our power to something or someone outside ourselves, or to summon the courage, integrity and impeccability to return home by walking the challenging, but ultimately enlightening Black Road of Spirit.

APPENDIX A: TESTIMONIALS

[The following unsolicited Testimonials from clients of the Regenetics Method are for educational purposes only and make no medical claims, promises, or guarantees. The first two Testimonials are by the same person writing soon after Potentiation Electromagnetic Repatterning then just after completing the 42-week process. The latter Testimonials focus on Articulation Bioenergy Enhancement, Elucidation Triune Activation, and Transcension Bioenergy Crystallization. Additional Testimonials to all phases of Regenetics are available online at **http://www.phoenixregenetics.org.**]

"It hasn't been quite two weeks since I received Potentiation and so many things have already changed. I was really losing ground before Potentiation and felt I would die from complications from pesticide exposure last summer. The night of my session I felt that toxic energy being removed. I was so deeply moved and uplifted I could hardly sleep. The next day brought waves of detoxification but nothing as challenging as having allergic reactions. That morning I had a giant breakthrough and experienced bliss throughout my body for hours. It felt like liquid light was surrounding and flooding my being. My body was peaceful and a new very, very quiet place inside me emerged. I'm sure this quiet place has always been there, but the noise of the war in my body has prevented me from hearing it. Before Potentiation at night I would start to lose motor coordination. Now I can get down the stairs even in the middle of the night without a problem. This is significant because in the past it literally took me hours in the morning to get on my feet. Also, the moles on my neck have faded and half of them are gone. Even my urinary functioning is back to normal where I used to have to strain. I simply cannot thank you enough." DM, Orcas Island, Washington

"I have completed my nine months for the Potentiation process and boy am I happy. I now have a life where before I had only restrictions. I can drive my car without wearing a mask. I can shovel shavings for my horse's stall without feeling dizzy and sick. I can stay on the computer 6 to 8 hours a day instead of less than one hour. I have a great boyfriend. I am well enough to work and earn a good living—something I've waited 15 years for. The person I was before Potentiation was so physically damaged by heavy metal and other toxicity it was just a matter of time before a nasty reaction would have sent me out with heart failure. It's difficult to describe in words, but I feel new and renewed, as if the best part of me expanded and everything else, including my brain fog, just disappeared. My heartfelt thanks." DM, Orcas Island, Washington

"Since Potentiation I generally have a sense of greater wellbeing, stronger workouts, less sugar and food cravings. I seem to be taking better care of myself, extending myself a certain tenderness, suffering less anxiety. It feels good!" CE, Asheville, North Carolina

"It's now 18 months after we first experienced Potentiation and I wanted to write about this first activation because in some ways it is the most important to me. My husband had been horribly sick with Chronic Fatigue (CFS) and fibromyalgia for nearly 3 years. We had tried countless therapies and moved to Hawaii from New York City in hopes that his 32-year-old body would recover. Every day was just pain for him; he couldn't think, had bad anxiety, heart palpitations, constant nausea, and couldn't sleep for more than 2 hours at night. We signed up for Potentiation. He felt visibly improved in the days right after. I was so happy! A week after Potentiation he was at his computer reading and interested in work again. He could think! The anxiety stopped. In the weeks that followed, he was able to get more and more sleep. As I'm writing this testimonial 18 months later, after completing Articulation and Elucidation, he is pain free. He sleeps like a log at night. He works out three days a week at the gym. While going through the DNA activation, we continued exploring ways to heal the physical body. There are many things that helped the detoxification process and some that didn't. In the end, many things contributed to my husband's recovery. Potentiation, however, in my mind, truly commenced the healing journey as well as provided a timeline that brought us both back from darkness." EK, Kamuela, Hawaii

"I am so grateful for Potentiation. It has really helped me turn the corner on emotional reactivity in important relationships. In the midst of conflict I am able to both understand the other person and have on-the-spot clarity about my feelings and the needs of the situation, and communicate it, with respect for myself and the other person. It has transformed conflict around here, with lovely effects that have radiated into my extended family, my daughter, my son and my mother." JS, Colorado Springs, Colorado

"I'm in my third month of Potentiation and I wanted to let you know the profound effects it has had on my life already. First of all, I've kept a Potentiation journal, which I feel has helped me maintain positive intention. I've been doing a detox program with my alternative healthcare professional and was aware of the beneficial effects of Potentiation on that program. But what really made it concrete was when I recently did an ionized footbath. I put my hands and feet in a tub of warm water with an ion machine. My practitioner could identify the waste being pulled from my lymph and joints, including heavy metals. What he saw made a believer of him. He said he'd never seen anyone draw out so much 'stuff' so quickly. I could even feel toxicity being pulled out of my brain. It was truly amazing. I'm looking forward to Articulation at the five-month mark, followed by Elucidation. I feel very blessed to work with you both. I feel and know your work is truly transformational." MR, Assumption, Illinois

"The past 5 months since my Potentiation have been amazing. What a process you have developed! I noticed the following ... I went on a HUGE house cleaning binge. I was tired but somehow had this incredible energy to clean everything in sight—like never before. I couldn't believe it. Later, a bump that I had above my eyebrow got itchy and burst open. I'd had it for over 15 years and thought it was just a part of me, but then it completely disappeared, healing without any scarring in just 3 days! Wow! I've also noticed that I have better tolerance for foods. My daughter has also greatly improved on the foods she can tolerate and her digestion. Finally, I've had an inner confidence that I never had before and am starting a private practice after hating my office job for many years. I'm really looking forward to the rest of the process to see where this will lead me!" AC, Ontario, Canada

"Thank you so much for last week's Potentiation session. I've been involved with energy work for a long time and have never experienced the kind of shift that has happened since connecting with you. I've already released a lot of emotional garbage. The amazing thing is that even though it was intense, I felt supported. Physically, in fact, I feel better than I have in years. Since the onset of my chronic environmental illness many years ago, any strong emotion would rattle my nervous system and make my allergies worse. Not now. I feel good. The emotional stuff has cleared. I never really expected to feel completely well again, and it's a little mind-boggling thinking about what I'm going to do with my life as a fully functioning person." SP, Orcas Island, Washington

"A subtle but very real thing occurred just three weeks after Potentiation. I'd suffered a math phobia for most of my life, ever since having my head thrown at a blackboard for getting a math problem wrong in grade school. (Prior to that, I was a math whiz!) So severe was the trauma that I cringed at anything that had to do with numbers. Years later, after Potentiation, it felt like tiny tiles were releasing from my brain and suddenly the calculator became my friend as I looked forward to numbers and calculations, and even to doing the financials in clients' business plans that I'd formerly farmed out to accountants. This was so exciting and freeing! Over time, my allergies and asthma also faded into a distant memory. The suffering was simply gone, and more and more vibrant health was becoming mine. As I went on to Articulation and then Elucidation, I experienced greater depths of living and loving. The separation I'd always felt changed to a oneness with everything and everyone. New possibilities and fresh insights arose. My entire being changed to a 'Light-ness' I'd never felt before. As an Interfaith Minister, counselor and coach, I've learned and shared many modalities to effect changes in clients. Some changes 'stuck' longer than others, but none ever deeply and profoundly changed the client at the core level and lasted like the transformations I've personally experienced through Regenetics DNA activations. This one modality directly and permanently promotes transformation at the physical, mental, emotional and spiritual levels. I am forever grateful I found Regenetics." MS, Wood-Ridge, New Jersey

"At the time of my Potentiation seven months ago, I was depressed and suicidal due to my inability to sleep. I'd often go whole nights and only get an hour of sleep. I still struggle somewhat with insomnia, but I no longer, or rarely, have the kind of nights I used to have before Potentiation. I've also suffered for years from TMJ, but in spite of braces and a mouth full of metal, I've improved in this area as well. I still have many challenges, but I now generally believe life is worth living and I even have feelings of happiness and joy. I also had a serious case of Restless Leg Syndrome when I began this process. I still have it at times, but lately it has been a lot less. I know this doesn't sound like much, but trust me, it's a big deal. I think of all the many things I've tried, Potentiation has probably been the best investment in my overall health and has helped me the most." KC, Atlanta, Georgia

"I want to tell you about the subtle changes that have been taking place in my life since my Potentiation last week. I'm experiencing an overall calmness and wisdom. My loved ones seem more drawn to be with me. I'm communicating in a more specific and less emotional manner, and have started to just enjoy my time, no matter what I'm doing. I've started observing what 'is' and noticing what does and doesn't work, and am working on re-creating the life I want based purely on my potential. My health is improving every day; I'm definitely more aware of what is and isn't contributing to my wellbeing. It has only been a week since my session, and I realize I have so much more to be, do and have!" JH, Raynham, Massachusetts

"For years I looked for my Divine Purpose, which led me to begin studying natural healing. I did courses in quantum physics and discovered the Regenetics Method—and my whole world changed. I deeply appreciate that Potentiation and Articulation—to say nothing of Elucidation—have taken me into a new level in my own consciousness. This has been a gradual process of letting go of my old beliefs, my 'not being good enough' identity, and being reborn into a life that is exactly what I always dreamed of. Unbelievable blessings have happened in my life since I began the Regenetics Method. I now have abundant energy, health, and wealth! The Regenetics activations have supported my spiritual growth and evolution in a relatively easy way. My body is no longer a burden, but more energized than ever. My energy sustains over time in a new way. I realize that I'm a creative Being with increasing amounts of blessing, joy and delight in my life. My finances have improved dramatically, as have my relationships. I find myself being able to help people as never before, with access to the right words at the right times, while no longer being 'caught' by my emotions as before. Finally, amazing 'coincidences' have become increasingly evident and common in my daily existence." AV, Pretoria, South Africa

"I consider myself very open-minded and accept the power of spirit to heal physical and mental imbalances. Yet last year when I was suddenly confronted with severe food and chemical allergies after a trip to Haiti, I felt humbled by limitations I'd never personally experienced. A yoga and meditation practitioner for many years, I thought I was immune to chronic physical ailments. A close friend recommended Potentiation, and since I'd noticed significant improvements in her, I decided to try it. I prepared myself mentally for a week, drawing on my own knowledge of the power of intention to heal. My Potentiation session itself wasn't at first very different from deep meditation. Soon, however, the results were astonishing. In less than two days I felt my allergies completely leave my body, as well as mucus-forming food 'intolerances' to wheat and dairy that had plagued me since childhood. I could eat anything, though I'm still vegetarian for moral reasons, and could finally breathe through my nose again! I strongly recommend this healing process, but be very pure and strong in your intention as to why you're 'potentiating.' Don't forget it's your higher mind you're connecting with, and that healing starts in you." DR, Marshall, North Carolina

"Two years ago I was frantically looking for healing for a degenerative eye disease that was slowly blinding me. I totally rejected the idea of an operation, the cure offered by Western medicine. After reading *Conscious Healing*, I knew that I must try the Regenetics Method. During the first phase of this method, Potentiation, I fell into a dream where I was a giant stick figure striding through the universe. One or two steps was enough to take me to new Suns. Over the next five months, I was aware of a wave of energy slowly working its way through my energy fields and physical body. I experienced a slow, usually very gentle process in which I gained a newfound appreciation of life. Fascinatingly, Potentiation helped me overcome my eye disease—though indirectly. In a subtle but extremely important way, this work encouraged me to move beyond my duality and its resultant 'blockage' where the Western medical approach to this particular issue was concerned. Also, I believe that the enlivening process I experienced as Source energy tracking through my energy body after Potentiation helped prepare me on many levels to accept perfect success from my eye operation, when I finally got around to it. I most highly recommend Potentiation as well as the rest of the Regenetics Method to anyone who is looking for beneficial change in a variety of areas related to physical disease or pain, including discomfort and issues rooted in emotional or mental blockages. If you wish to have a greater conscious experience of yourself as a spiritual physical being, the Regenetics Method is definitely for you." DM, Montreal, Canada

"Thank you for a most graceful entrance into my Potentiation process. I shared the experience with two other people, one receiving Potentiation and one who was seven months into it, and found that the combined energy and intentional bond we formed brought us all to a state of ecstasy. When the session ended, nothing but radiant bliss was pulsing through my body. I felt moved to embrace the person I was sharing the experience with and we both felt the energy magnify. I've been in the holistic field for many years, and Potentiation has been one of the most profound processes I've ever experienced." JC, Novato, California

"I'm definitely feeling a shift since my Potentiation two weeks ago. Many 'stuck wheels' have begun to move again in my life. I feel strangely 'uplifted' from inside out, as if something almost structural is being built. This is accompanied by a sense of stability and support I've never felt before. Prior to my session I felt I was sort of 'crumbling' inside, but now I feel I'm 'rebuilding.' This change is very obvious. Emotionally, I'm doing better than I've done in a very long time. Many issues I've struggled to understand for years have suddenly been made clear to me. I've also gained a more profound awareness of Christ presence through this work. I've worked with the best energy/spiritual healers on the planet over thirty years, and Potentiation definitely rates among them—with the distinct difference that this also feels like something tangible is being constructed. Potentiation is certainly the 'reset' that you call it." LH, Black Mountain, North Carolina

"I began to feel a strong vibratory connection with this work when I read *Conscious Healing*. I read it through the first time in one day, and though I didn't take everything in, I could 'feel' that this was the right thing. The next day I contacted the Developers to set up an appointment. From that day the process began for me. My posture easily changed and became more upright. I began to feel a deep calmness and serenity. I felt enormously confident that this was going to make a huge difference in my life. Throughout the day of my Potentiation, I felt as though I was being 'worked on' in different energy fields. An hour before the session I felt a great surge of joyful energy. I have since read your book several times and am so thrilled to learn about some of the science to back up what I suspected and have been reading about in other texts. Your courage to put some very controversial texts together with the science was so gratifying for me. Your writing and the way you explain and connect things put a lot of disparate pieces of so many puzzles together. I am also grateful for your generosity to tell your story about your own bodies. I had been suffering from many physical challenges and feel absolutely certain that I have turned a corner and that these things are now part of my past. Up until the Potentiation session, I had been struggling with the most vile and crazy-making rash. The only thing that would relieve me of the severe discomfort was an over-the-counter antihistamine that left me feeling lousier than I did with just the rash. After the session I just knew that I could stop taking the drugs, so I went to bed drug-free for the first time in months. I woke up feeling refreshed with no itching at all. If this is all I have to look forward to, it would be enough. I know this is just the beginning, though. I give thanks for all the ways you were guided to facilitate this work and even give thanks for all the years of symptoms that led me to try this process, because it is truly a blessing to witness and be the recipient of such vision and revelation." DP, New York, New York

"Since initiating the Potentiation process 4 months ago, I've experienced several rather profound changes. My hot flashes have subsided, which is a real relief and allows me to sleep better at night. Most of the arthritis in my hands and knees has cleared up, reducing my dependence on glucosamine. I've gone through some detoxification and

can sense more energy coming in through my chakras."
PD, Myrtle Beach, South Carolina

"What do I feel is different since last week's Potentiation? There will be a big layoff soon where I work, but I feel much lighter, calmer and more positive. A lot of my depression is gone. Also, for the last couple years I've felt the left side of my body tightening and I've constantly had the urge to stretch it. One doctor said I had Illio Tibial Syndrome and gave me stretches to do. A massage therapist/instructor said I had a fascia problem. Being treated by them helped very little. I underwent other unusual therapies, also with little results. But now I've had a major shift and tremendous release on my left side, especially in the left hip and leg. I really cannot believe it. My hip is so loose, lighter feeling and rounded out. I do massage therapy and for years have worked on myself. But now there is no need to work on these areas!" CH, Streamwood, Illinois

"My experience of Potentiation was both subtle and powerful. During the early phase, the first few months, I felt an unusual sense of happiness and peace and an overall subtle shift inside. Then, as the process unfolded, I realized my food allergies had completely disappeared! I'd tried other treatments with limited success, but with Potentiation I gradually noticed I could enjoy food that would have normally caused headaches and spaciness. Very remarkable!" EL, Asheville, North Carolina

"My personal experience of Potentiation was powerful, initiating a great deal of emotional as well as physical (pain) release, especially in my head—which as a massage therapist, I found most interesting! As a child I had a lot of head injuries and also contracted meningitis at two and a half. Throughout my adult life, in therapy and elsewhere, I'd attempted to get to the core of the abandonment issues around my traumatic experience of meningitis (when I was quarantined away from my parents for days), but on some level I knew my cells were 'hanging on' to the memory. The very day of the Potentiation process when I entered the seventh electromagnetic field, which in my Electromagnetic Group regulates the musculoskeletal system as well as many emotions associated with abandonment, I was massaging a client. Halfway through I started sobbing and couldn't stop. I felt exactly the same sensation I felt during my meningitis episode: that of being catapulted into space, disconnected from everybody and everything, floating in a dark tunnel. Until Potentiation it wasn't possible to feel, at a body level, the full emotional impact of this experience. But now I was feeling it. Somehow I finished the massage and called two friends to be with me as I underwent the most extraordinary release over twenty-four hours during which I felt my cells were literally being cleansed biochemically of the hormonal residue of my childhood trauma. Within two days my lifelong fear of abandonment was merely a memory." SL, London, United Kingdom

"Since my Potentiation I feel great! I have loads of energy and generally feel balanced. During and just prior to my session last week, I felt immensely relaxed, deeply tranquil. This afternoon I plan to go jogging. I have strength that I haven't felt in quite some time. Thank you so much for sharing your healing work with me and humanity." AA, Asheville, North Carolina

"Even though I'd been on a journey of healing and conscious mastery since 1977, and had made progress in becoming integrated and embodied, and had known periods of great joy and profound transformation, I still felt that a fundamental part of me would never catch up with my mental understanding of such higher states of being and awareness. Though I was in many ways more satisfied and recognized than ever before, I still felt somehow 'fragmented.' At 59 as a professional therapist, part of me felt myself beginning to die inside, tired of it all, my life force waning. As soon as I discovered the Regenetics Method and read about the Fragmentary Body, the missing piece for me, I contacted the Developers and booked my initial session. By supporting me through the release of two limiting relationships and the 'upgrading' of several more to healthier states, Potentiation helped me realize something I'd been unable to achieve before in all my spiritual and therapeutic pursuits: a sealed energy field with no more ongoing energy loss and no more vulnerability to compromising situations and relationships. In my estimation no amount of therapy or mental processing could have achieved comparable results. I actually felt the sealing of my energy field and, to this day, have no more problems navigating challenging circumstances, personal or professional. Regenetics has tremendously improved my work as a therapist as well as my personal relationships, old and new. As a fringe benefit, I'm often told I look ten years younger! Certainly, I feel younger, excited to be alive again, with consistently more joy." AW, London, United Kingdom

"Thanks for the wonderful Potentiation session last evening. I felt an armor (with lock and key) lift off my heart area with lots of emotional release. The emotion of FRUSTRATION kept coming up along with a HUGE wave of energy starting with my head, heart, stomach and root chakra areas. I am very grateful for your wonderful healing abilities." CS, Atlanta, Georgia

"Thank you so, so much for all your help! You've truly been a blessing from God. The day after Potentiation, my chronic rash disappeared 99.99%. The only thing that seemed to hang with me was a few small spots that only had an irritated sensation on an occasional basis. I feel that these spots are only energy vents, clearing out old residue. All in all I feel GREAT! The positive energy I feel has continually gone in a forward healing mode, which has been so exciting. THANK YOU SO MUCH!" LM, Sylva, North Carolina

"My daughter was so pleased with her Potentiation results I had to try it. I had severe food allergies that had escalated through the years to the point where I could barely eat anything without discomfort. Within the first month after Potentiation, I found myself able to add more foods to my diet. It has now been two months since my session, and I feel I've improved at least 70%. I'm 82 years old and to see my health improve so fast is thrilling." LH, Sylva, North Carolina

"It is almost bizarre to retell something that now seems so distant, but I am inspired and guided to share this with others who may be searching and would cry tears of relief at hearing of another's triumph. In the late 90s I was living and working as a teacher in New Jersey. I started noticing lowered immunity, depression, frustration with personal relationships, and a deep ache inside to 'feel good.' I began listening to a radio talk show about health and attending 12 step meetings and Unity Church in search of a daily way of 'connecting.' I resigned from my position as a middle school teacher and began my studies in Nutrition, Natural Cookery, and Naturopathy. Upon returning home to Virginia in 2000, I crashed. Suddenly I was having serious trouble sleeping, became severely bloated, had major digestive discomfort and distress, felt I would die each month before my period, experienced anxiety, depression, 'brain fog,' malaise, fatigue, aching muscles, fear, and despair. I experimented with practically every diet known to humankind, in addition to energy work, acupuncture, allergy elimination techniques, EFT, hardcore supplementation, sauna therapy, reiki, electrical cellular stimulation, IV therapy, reflexology, psychic detox techniques, acupressure techniques for emotional release, massage, heavy metal detox, German machinery, microscopy, and others. Although some alleviated my symptoms, I still did not feel a fundamental shift. One morning at 4 am, I was guided to do Internet research and found a commentary on a bulletin board about someone's boyfriend who had 'taken Potentiation' and healed his Leaky Gut Syndrome. Thinking Potentiation was a bottle of homeopathic drops, I contacted Sol and Leigh and began my journey through the Regenetics Method. The level of neuro-toxicity (caused by vaccine damage and compounded by other factors) I carried was quite high and Potentiation was the perfect avenue for its departure from my body. The process was intense at times, but I could gradually feel a lifting of the 'sludge' and knew I was headed for a life-affirming existence after years of extreme discomfort. This technique has helped transformed my life beyond anything I could have imagined." CH, Dallas, Texas

"My Potentiation session was a transformative experience. I felt a deep connection, sort of the way one can feel an acupuncture needle activating a meridian point. I also felt, and continue to feel, a kind of turning of the mind away from negativity, an enhanced ability to move out of, or not fall into, what I call 'bowling alley gutter mind.'" ET, North Hampton, Massachusetts

"I can't believe it's been nine months already since Potentiation. There have been a lot of changes in my life, internally and externally. I'm starting a new career, and my health's never been better. But most importantly, my consciousness has gone through quite an overhaul. Life and the universe look totally different now, and I enjoy much more freedom and gratitude than I ever imagined before. I seem to encounter friends, teachers and inspirational materials on a constant basis. There is a strong sense that finally everything is coming together. Thank you so very much for the service you provide." LB, Little Rock, Arkansas

"Since my session I've felt wonderful. I've been trying several therapies, but I know Potentiation has been the catalyst for any real healing. I just attended an amazing lecture by Dr. Len Horowitz, who explained how DNA is really just an antenna to receive love and light from the Creator, that all healing comes from these higher energies, and that when our DNA gets clouded with chemicals, toxicity and negativity, it shuts off our connection with the universe and ability to heal ourselves." SM, Denver, Colorado

"I was not very skeptical about the concept of DNA activation prior to the Regenetics Method. I had heard of other methods, but was waiting for yours. There is no doubt it worked for me. While I have not fully gotten over all of my environmental allergies, I was pretty bad, and have much less attachment to them now. Rare is the day now that I wake up not breathing through my nose. Lately, I eat pretty normally, for me that is, and have virtually gotten over sugar cravings. I have a better understanding of my purpose, and more energy to carry it out. I had the most kundalini movement and most intense experience of awareness of everything around me from Potentiation; but I think that was because it initiated the most change in my case. I still have some back pain, but it is a lot better and it does not have the emotional impact it used to. A big step has been taken toward realizing my intention for the activations. Full fruition has not come; but I can see much of how it will. I am really rather pleased with my current manifestation, even though that might seem odd to a logical mind. I believe Regenetics is worth more than you are charging. It is so clear that it works, and even the sequence and timeline are accurate. I will recommend the Regenetics Method to those I think can hear it." BW, Goshen, Ohio

"I must say that since my Potentiation over five months ago, I've become extremely clear and focused in a way that I haven't been before. Very frequently my body feels so light and erect—I feel as though I'm pure light in movement—can't quite describe it in adequate words. I just know that I wouldn't trade how I experience the world within for anything. As part of my experience I've literally felt torsion shifts—almost as though a chiropractor is working on me. Thank you." GP, Toronto, Canada

"Nearly three months into Potentiation, the love I feel in my heart is divine and I am really embracing the beloved in me like never before. I feel amazing, my body is splendid, I am eating many more foods (though still not doing great with too much sugar), and my energy is fabulous. My leadership has amplified considerably, affecting all the men and women I am in vibrational union with throughout the world. I am generating much more in all areas, and my mother and I are working together in a new way. I feel happy and joyful—like I have moved into a whole new place of being and am really embracing my natural gifts and developing them at a deeper level. Some people are leaving my life, but most importantly, I am no longer triggered by the duality in others and am maintaining my unity at all levels. So thanks to you! Your work is amazing. God bless!" SF, Sydney, Australia

"Thank you so much for my Potentiation session yesterday. I definitely connected with you and it felt very powerful. While meditating during the session, I experienced tingling feelings which continue at times today—waves of energy moving through me, a very warm and expansive feeling in my chest/heart chakra area, and definite feelings of joy and an uplifted sense. I also starting feeling shifts prior to the session after reading your book, which is fascinating and inspiring. So much of it resonates with me. This may sound strange, but it feels like I'm absorbing essential nutrients as I take in your words. It is amazing work that you're facilitating. I already feel different: lighter, more clear-headed, energetic, and hopeful. I've had carb hangovers whenever I had even a little bit of carbs or sugars, and have thus had to totally avoid them for years, severely limiting my ability to eat out/socialize/travel. I do a lot of self muscle-testing and while I know it's early in my Potentiation, I'm AMAZED that I already test strong to healthy (organic) carbs and even straight sugar! Wow! As well, chemical sensitivities have dominated my life for 18 years. I'm already feeling shifts here as well and will keep you posted on my healing. Eliminating MCS and food intolerances will create a quantum leap of improvement in all areas of my life. Thank you again. I will indeed spread the word to all those who can hear it!" JS, Cumming, Georgia

"Many thanks for being midwives again in the Articulation phase. I felt perfectly connected to my source: energy pouring and flowing. I feel more in balance than ever before, equanimity is probably the best word to put it together. Mindfulness is another aspect I experience deeper and over longer periods. Being aware that I am the one consciousness. Thanks again." HD, Echteld, Netherlands

"Following my Articulation I'm experiencing huge waves of energy. Additionally, I've had big energetic openings that relate to the second and fourth chakras—physical energies shifting as well as beliefs and emotions. I'm definitely moving into more of a 'third-person' awareness, which is kind of a surprise in the midst of all these powerful awakenings. I've also had lots of mostly positive movement in the relationship area. Strangely enough, with all of this going on I feel I'm definitely moving to a place of internal stability and balance. This is truly some powerful work!" TV, Cincinnati, Ohio

"I felt and continue to feel the Articulation more profoundly and consciously than the Potentiation. It feels very deep but also gentle and nonintrusive. I have been meditating for 13 years and have historically found it difficult to carry the consciousness of my meditative states into my everyday life. So I have often felt like two people. Since Articulation I feel more present, more incarnated, and more like one person. Interestingly, my son developed a fever within a few hours of the Articulation. He was asleep by the time of our session and I didn't check on him until much later, so I don't know if he developed the fever immediately. But he looks wonderful and lighter, as if some burden a five-year-old should not have has been lifted from him. Something wonderful is happening." AM, Phoenixville, Pennsylvania

"I woke up this morning following my Articulation very late, having had a very deep sleep! You two deserve the Nobel prize for the work you do! Let's just say my body is 'getting rid of' what made me so ill in the first place. I can actually see my skin dumping bacteria and pathogens. It was easy to connect with you both at the Theta frequency! As one of my heroes, Richard Bandler, says: All healing is faith healing. Last night I just kept saying to myself, 'I align with Sol and Leigh,' and I could feel the whole process work: audio, visual and kinesthetic! For me after only 12 hours, the results have been staggering, to say the least!" DS, London, United Kingdom

"Since our Articulation my daughter is quieter and I'm more relaxed. In fact, I've been in an almost constant relaxed state ever since. Even when provoked by a huge argument an hour after our session. I have to say I feel good. A longer, deeper 'high' than with any other work I've ever done. Feeling elevated from energy work longer than a day is pretty impressive, if you ask me. My daughter says, 'My skin looks clearer and I don't feel so stressed.' While she seems more lively, I feel more caring and less inclined to worry. So double yay!" MH, Asheville, North Carolina

"I had Articulation done about two weeks ago—have waited to write to see how things go. Think I've lost 1 or 2 allergies—one to lemons and one to peanut butter. I feel pretty good—lots of energy and a positive attitude. Find myself celebrating just being alive—not very usual for me. Thanks for your amazing work." DB, Juneau, Alaska

"I have experienced much shifting since the Articulation session. I am releasing energies and experiencing much more 'happy' states and balanced states. Feelings I have never, in this lifetime's memory, experienced before. I have seen the perfect atonement since the session and have much to look forward to in continuing to work with you and your sessions. I am already excited to be in contact with you for the next session when that time calls. Your beautiful work and sharing of your gifts is a blessing to all." JS, Hartford, Connecticut

"I did indeed feel the energy strongly during my (our) Elucidation session. It was different this time from previous sessions. I felt a very high, intense energy around my head, but only from the ears up, as if my head were divided in half horizontally. It felt as if there must be a very high vibration being directed to that part of my body. Below my ears/nose to about midway on my chest was a more mellow energy. I really didn't feel anything further down on my torso and legs. During the session, the 'above' energy lessened a bit in intensity and seemed to almost meld with the 'below' energy, but the feeling of having a 'divided head' remained distinct. I found it interesting and thrilling to experience. Thank you both for being such wonderful Lightworkers, and for creating, developing and facilitating this amazing healing process." EL, Daly City, California

"I have been going through the most mind-blowing transformational experiences ever, and I just checked up on how it corresponds to the unfoldment of Elucidation on the Schematic, and I have goosebumps right now! It is all too complicated to properly convey, but I at least wanted to say to you how much I appreciate the work you are doing and the gift you are providing to humanity from the bottom of my heart. Thank you so much!" CR, Vancouver, Canada

"Right after my Elucidation session, I felt so light and clear. I felt blocked energy releasing right away. My kundalini turned on and I experienced incredible warmth emanating from my second chakra area and then throughout the rest of my body. I noticed three distinct shifts in my energy. My emotions went from frustration and pain to enthusiasm and joy. It feels like a great burden has been lifted from my space and I don't have to struggle anymore. This has been the most dramatic of my sessions as far as my noticing energy shifts and changes during the session. I expect the months ahead will be interesting! Thanks so much for doing the work you are doing. It will definitely be one of the primary healing modes of the future." JE, Oakland, California

"I was introduced to *Conscious Healing* through another DNA activation certification program I was being trained in, which assigned Book One on the Regenetics Method as required reading. The first time I read *Conscious Healing* I felt a strong connection and a feeling of unity and resonance to the material and the creators of this work. I started the Regenetics Method one year ago and just recently entered the third Phase, Elucidation Triune Activation, which was the most powerful session so far for me. I have been on a conscious path of healing and wholeness for two decades, and it is extraordinary to experience the deep healing and transformation that is taking place within me through the Regenetics Method. Deep-seated, unresolved emotional issues are being surfaced and cleared through the torsion energy now filling my bioenergetic field. I am able to observe the process while simultaneously experiencing an expansion of consciousness which is giving me a feeling of spaciousness and freedom in my awareness. Through this work I have gained a greater sense of being the creator of my life. I am very clear that I have freedom of choice in areas that once felt difficult and stagnant to deal with— regardless of how much inner work I did. Especially around the area of money and finances, I used to feel stuck. It is just in the last five months of this process that I am experiencing substantial movement and have become a partner in a business that is flourishing and thriving! I have also experienced healing in my physical body. Before Potentiation I was making frequent trips to the bathroom in the middle of the night. An acupuncturist told me my kidneys were deteriorating. Well, my kidneys are stronger and healthier than ever, and for months now I have enjoyed a good night's rest. I highly recommend the Regenetics Method. It is effective and efficient, and is also a great value!" LB, Los Angeles, California

"Thank you so very much for your wonderful information, feedback and everything that you do, which is such a beautiful gift to the world. I was so thrilled to read your latest news, with the latest addition to your Method. I have already signed up for Transcension and look forward to hearing from you. I cannot even mention to you what has happened to me for the last two years ... It is still going on. I can now take control of any situation and have learned the real meaning behind 'unity consciousness,' which is the love one feels for everything and everyone ... no matter what harm is done to you. I can assure you that nothing has helped more than receiving your help through the Regenetics Method!" PH, London, United Kingdom

"I have just completed the fourth activation of the Regenetics Method, Transcension, and immediately I had a totally new perception of Oneness, in which I simply am all the things around me, even the whole world ... indeed, all that is. Before it was just a concept; now it's reality. Funny, really, that I didn't see it that way before. It just makes you smile and feel very, very serene. The activations that my wife and I received have greatly increased the occurrence of synchronicities as we work to achieve our life purpose in collaboration with others, on a scale not considered possible before." PS, Neuilly, France

APPENDIX B:
FREQUENTLY ASKED QUESTIONS

Q: What distinguishes the Regenetics Method from other DNA activation techniques?
A: Those with highly evolved consciousness such as spiritual teachers always have insisted the body-mind-spirit can be "potentiated" by words in the form of songs, poems, prayers, affirmations, or mantras. The sound of the words must be harmonically attuned to the organism and the intention behind them impeccable. This is why although DNA activation has become trendy, results can vary enormously—ranging from none to life-changing. The more advanced the facilitator's consciousness, the less need there is for mechanical devices. Some shamanic healers believe the digital recording is like a clone—lacking spirit—which calls into question the effectiveness of DNA activation CDs and MP3s. This touches on a discussion of the Path of Technology (giving away one's power to a technological intermediary) vs. the Path of Nature (self-empowerment or conscious personal mastery), but the bottom line is: there is no substitute for live human consciousness and voice. In addition, we are aware of no other DNA activation modality specifically designed to seal and heal the Fragmentary Body. This is a key point because without bridging duality at the level of the bioenergy fields, it is impossible to build a higher energy body. We have performed Regenetics sessions for many individuals who have experienced prior DNA activation, and the typical response has been that what we do is uniquely powerful and effective. Finally, Regenetics employs the recently rediscovered Solfeggio scale. The Solfeggio is a primordial six-note scale many scholars believe to be the creational scale. This scale is so transformational it was hidden by the Roman Church for centuries. One of the Solfeggio notes, "Mi," is a frequency that has been used by molecular biologists to repair

genetic defects. This is also the note used in Potentiation Electromagnetic Repatterning.

Q: Do you suggest any resources or activities that may support the Regenetics Method?
A: Our personal experience and professional observation have been that drinking several quarts per day of pure water and eating organic food (including enough starch to bind toxins being pushed by torsion energy out of cells) tremendously support the process. For us, this was especially true after our nutritional allergies disappeared following Potentiation. In addition, light exercise such as rebounding, swimming or walking is an excellent way to keep the blood and lymph moving to increase oxygenation and assist detoxification; while getting enough quality sleep on a regular basis is deeply restorative. In addition, there is no substitute for engaging in activities that inspire feelings of creativity, joy and love, since genetic research clearly shows that such uplifting emotions fortify DNA. Finally, to access a wealth of helpful information and perspectives on all phases of the Regenetics Method, we encourage clients to join our private Regenetics Method Forum. For Member Guidelines, visit **http://www.phoenixregenetics.org**.

Q: Is it possible for me to "mess up" my Potentiation?
A: As long as you approach your Potentiation with an open mind and heart, we know of no way you can botch your DNA activation and electromagnetic repatterning. Note that this includes simultaneous or subsequent exposure to other forms of DNA activation, other types of energy work, and even environmental radiation sources such as computers and cell phones.

Q: Are there any things I might do that otherwise can interfere with my experience of the Regenetics Method?
A: Nothing except an individual's free will can impede the proper unfoldment of Regenetics activations. That said, there are some basic considerations for clients passing

through deeper periods of healing during Regenetics. Our advice always is to trust your intuition in supporting the unblocking of distortions in—and detoxification of—your physical, mental, emotional and spiritual bodies. If something feels as if it is adding too much "fuel to the fire," it probably is. Sometimes, supplemental healing activities are entirely appropriate. Often, however, especially for those learning to trust the wisdom of their body-mind-spirit as it self-corrects its imbalances during Regenetics, "less is more."

Q: Will the Regenetics Method interfere with any other energetic methods I might be trying?

A: To the contrary, the Regenetics Method may make other modalities more effective—and even too powerful at times. This applies not only to energetic therapies, but to any modality. Whether to try other modalities following Regenetics is entirely up to you. Always trust your intuition. You, and only you, know what is right for your body, mind, and spirit. Our only caveat is that you ask other practitioners to treat you very gingerly, as the sealing of the Fragmentary Body that occurs at approximately the five-month mark of Potentiation means that your bioenergetic system will accomplish more and more with less and less outside energy input.

Q: Does Potentiation change my basic DNA?

A: No. Potentiation Electromagnetic Repatterning simply activates a latent "program" in potential DNA designed to "reset" the human bioenergy field at a higher vibratory configuration we call an "infinity circuit" based on the number 8.

Q: Do I have to understand the Regenetics Method fully for it to work?

A: Emphatically not. The Regenetics Method is inherently an experiential, not intellectual, process. Parents have reported positive results in very young children, for example. Few people "understand" the medicines they try. That said, the more one commits to thinking in this new way about human potential and the ability to re-create ourselves at the "ener-genetic" level, the more one can engage through intention in the process of conscious personal mastery.

Q: Do you need a bloodspot or other personal artifact such as a photograph to activate my DNA?
A: No. Only your name, date of birth, location, permission and intention are required. The latter two are assumed on scheduling your session and receipt of your educational service fee. Mutual intention establishes an ener-genetic connection on the "biological Internet" constituted by DNA, allowing DNA activation to be "emailed" to the correct recipient. In the case of responsible adults, we never perform a Regenetics session without an individual's conscious permission.

Q: Should I have my aura cleared before experiencing DNA activation?
A: We are aware of DNA activation proponents who claim that the aura should be cleared prior to DNA activation. Given that it has been established that DNA projects and regulates the aura, not only does such a claim make little sense; it is impossible to clear the aura permanently without activating DNA. While DNA activation might be facilitated by subsequently clearing the aura, effective DNA activation is the primary modality and is capable, by itself, of clearing the aura over time.

Q: I've read online that DNA activation is dangerous. How do you respond to this?
A: Such thinking reflects programmed victim consciousness and the propagandistic view that the world is an inherently "dangerous" place. The simple fact is that listening to beautiful music or making love stimulates DNA. Our genetic structure also is activated continuously by cosmic gamma rays. DNA activation has been truly a god-send for us, the individuals we have trained as Facilitators, and hundreds of clients—many of whom had given up hope of ever being well again prior to discovering the Regenetics Method.

Q: Are there any contraindications involved with the Regenetics Method?
A: As for contraindications, there are no "indications." Regenetics is not a therapy, but a Method of facilitating conscious personal mastery as a bio-spiritual healing path or way of becoming genuinely "whole." While making no medical claims, we suggest there is every reason to believe, based on hard science, that a successful reset of the electromagnetic disharmonies that have created problems can have a profoundly helpful impact.

Q: Do you need to know my symptoms, medical diagnoses or other issues to activate my DNA?
A: Symptoms, medical diagnoses and similar issues are unnecessary "baggage" on our end, as we endeavor to perform our DNA activations without attachment to a specific healing agenda. For a thorough explanation of our rationale, see our discussion in Chapter One of the enhanced effectiveness of prayer in healing when there is no attachment to the outcome. What is important is that you clearly set your own goals, as you will be the one integrating the energies of Regenetics over the weeks and months following your session(s). By focusing on conscious personal mastery as a path of healing or "wholing," the Regenetics Method represents a purposeful shift away from the diagnostic model. Too often diagnosis oversimplifies complex processes while "locking in" a problem in the sufferer's mind so that it seems nothing can be done. By far the most important factor in determining the level of success of the Regenetics Method is the individual's degree of conscious intention to use these energies for mastery in walking one's highest path in life. We are not saying you must believe completely in the process for it to bear fruit, but we do insist your willingness to approach your ener-genetic unfoldment with an open mind and especially heart greatly influences your experience of Regenetics.

Q: If you don't know what's wrong with me, how can you help me?

A: One of the fundamental precepts behind the Regenetics Method is that all illness, whether "physiological" or "psychological," arises from bioenergetic disharmonies appearing in the body's auric or electromagnetic fields. Similar conclusions have been reached by a growing number of scientists, including UCLA professor Valerie Hunt (author of *Infinite Mind*) and physician Richard Gerber (author of *Vibrational Medicine*). Through kinesiological (muscle) testing, we have observed an extraordinary level of consistency of energy patterns in the electromagnetic fields of particular groups of people. Each Electromagnetic Group possesses a unique arrangement in its bioenergy blueprint that applies to all members (see Appendix C). This is an exciting discovery because it renders individualized diagnosis unnecessary. To potentiate a person, we simply use surrogate muscle testing to determine the Electromagnetic Group, for which the Schematic serves as an informational resource for the client, and apply the same generalized DNA activation that is good for all Electromagnetic Groups.

Q: Is there anything I can do on my end besides putting myself in a "co-creative" state to assist my DNA activation?
A: The single most important thing you can do to assist your DNA activation is to open your heart and operate with love in all areas of your life. Revolutionary genetic research by Glen Rein shows that feelings such as love and joy positively impact DNA so that the torsion energy of universal creative consciousness can activate the genome's extraordinary transformative abilities. On the other hand, feelings such as fear and anger harm DNA so that it is less available for activation through consciousness.

Q: Could you say a word about setting and following through on one's intention during Regenetics?
A: Your flexible, heart-based intention is extremely powerful and important in actualizing the energies of Regenetics—although you will receive them in any case because the simple gesture of scheduling a session is an

intention. We recommend that you take time before your session(s) to clarify your intention by specifying (preferably in writing) the areas where you seek improvement. During your session(s), focus on the areas you want to address and imagine "downloading" healing energies into the places that need them most. Your goal should be to vitalize those areas that will allow you to reach your full potential. Keeping a "Regenetics journal" in the weeks and months following your session(s) also is a good way to stay focused intentionally on your transformation. Even a notation in your calendar or planner when you experience a positive shift is enough to paint a helpful picture of your unfoldment. Prayer, meditation and visualization are also excellent ways to maintain intention. You can continue to clarify your intention on a daily basis, by journaling or otherwise, throughout the process. To the best of your ability, be sure to maintain an attitude of nonattachment relative to your intention so that you avoid restricting or deflecting the desired outcome. The trick, we have found, is to put out intentional energy with genuine feeling, then freely release it so what you desire can come back to you. It is also a good idea to remain open to serendipity and trust your intuition, as other modalities and opportunities that present themselves may assist your unfoldment.

Q: How are Regenetics sessions performed at a distance if they involve sound?

A: Sound in higher dimensions is a torsion wave that, in keeping with recent research in Russia by the Gariaev group, can be transmitted instantaneously across theoretically infinite distances via DNA. Aligning with biologist Rupert Sheldrake's Morphic Resonance theory, this research demonstrates that DNA constitutes a "biocomputer network" similar to the Internet that, being present anywhere, is simultaneously present everywhere—effectively doing away with distance. In *Reinventing Medicine*, Larry Dossey makes a very strong case for nonlocal approaches to healing, noting that many scientific "studies reveal that healing can be achieved at a distance by directing loving and compassionate thoughts, intentions, and prayers to others, who may even be unaware these efforts are being extended." Our decision to perform Regenetics remotely is based partly on convenience; it allows us to touch people's lives wherever they are in the world. In addition, remote healing invites one to "think outside the box" of what we have been taught about the body and realize that, ener-genetically, humans are unlimited.

Q: Does the Regenetics Method involve anything resembling witchcraft or voodoo?
A: Regenetics is a technique for activating DNA. Many of our clients are deeply religious. The Regenetics Method is based on many accepted and emerging scientific theories relative to hyperdimensional sound and intention, nonlocalized mind, and the ability to activate the self-repair potential in "junk" DNA. Our research has focused specifically on the impact of sound and intention on this apparently noncoding portion of DNA that until recently has been considered useless. For some enlightening new takes on "junk" DNA, which we propose renaming potential DNA, see "The Unseen Genome" in the November 2003 issue of *Scientific American*, as well as Peter Gariaev's article "A Brief Introduction to Wave-genetics: Scope and Possibilities" in *DNA Monthly*. Our approach has involved kinesiology (muscle testing) to "map" the structure of the human electromagnetic fields. For more information on kinesiology, see David Hawkins' famous study, *Power vs. Force*.

Q: How does DNA relate to the body's electromagnetic fields?
A: The latest electrogenetic research likens DNA to a biocomputer that holographically interfaces with the human bioenergy fields, which in turn regulate cellular metabolism and replication. This new research flies in the face of traditional molecular biology dogma that considers DNA merely a biochemical protein-assembly code. In an article entitled "From Helix to Hologram" appearing in the September-October 2003 issue of *Nexus* and republished in *DNA Monthly*, longtime genetics researchers Iona

Miller and Richard Alan Miller write, "Life is fundamentally electromagnetic rather than chemical, the DNA blueprint functioning as a biohologram which serves as a guiding matrix for organizing physical form." The Gariaev group has shown that one can use linguistically encoded radio and light waves, or sound combined with intention (words), to activate DNA, which then can modify the human electromagnetic fields. These fields, in turn, are capable of modifying how cells are made and function. This noninvasive approach that represents the exciting confluence of energy medicine and molecular biology has been called "wave-genetics," of which the Regenetics Method is a human-potential-based application.

Q: Why in your opinion do traditional energy clearing techniques often fail to produce lasting results?
A: Traditional energy clearings work through the nervous and meridian systems as opposed to DNA. But geneticists have begun to refer to DNA, not the nervous system, as our biocomputer. In order to reset the human bioenergy blueprint and restore it to harmonic functioning, it is necessary to go directly to the root of the malfunction— which only can be accomplished via the genetic code. To do this noninvasively, one can employ hyperdimensional sound and intention (torsion waves) to activate the self-healing mechanism in potential DNA. Potential DNA, then, can revitalize the electromagnetic fields, and the electromagnetic fields, in turn, are capable of revitalizing the organism.

Q: Is there any twenty-four-hour avoidance period of foods or other substances following Potentiation as there is with NAET and its derivatives?
A: None.

Q: Does the Regenetics Method work like radionics?
A: Although the Regenetics Method was inspired partly by certain aspects of radionics, the remote energy

transmission used in Regenetics should not be confused with radionics. Instead of frequencies broadcast through the "psi-field" by way of a mechanical instrument to the client's nervous system, Regenetics employs particular vocalized sounds combined with non-directed (i.e., having no self-limiting agenda) healing intentions that work together to transmit an "energized narrative" via the morphogenetic Internet constituted by DNA. Radionics (like reiki and many other types of energy healing) is also mainly a light-based technology, whereas the primary energy of Regenetics is sound.

Q: I come from a homeopathic background and wonder if Potentiation addresses miasms?
A: Miasms, a focal concept in traditional homeopathy, might be thought of as paraphysical disease potentials latent in humans that are exploitable through toxic, nutritive, genetic, mental, emotional and even spiritual manipulation of the essential structures (nucleotides) of DNA and RNA. Barbara Hand Clow describes miasms as "etheric masses that hold memory of genetic or past-life diseases that were not cleared due to vaccinations, which prevented ... manifesting the disease memory and erasing it; or memory of disease [driven deeply into the body] by ... antibiotics, chemicals or radiation." Potentiation is designed, through DNA activation and electromagnetic repatterning, to begin the process of transmuting this negative "karma" and revitalizing areas damaged by miasms.

Q: In reading your materials, I see that you consider the emotional subtle body more primary than the mental subtle body. This is in contrast to other systems which say it's the other way around. How did you come to this opinion?
A: The emotional body in its purest, "transdimensional" form is higher than the mental body because before there can be spirit's descent into thought and resultant manifestation, there first must be the feeling of what that manifestation is to be. Thus the spiritual body is a form of pure awareness; the emotional body is the feeling or

desiring to extend that awareness into manifestation; and the mental body is where manifestation is constructed as it blends into the so-called physical. In the human "multidimensional" blueprint, however, as we indicate in our Electromagnetic Schematics (Appendix C), the mental body does appear to be higher than the emotional body. The former appears to be related to the sixth and seventh *chakras*, while the latter is related to the fourth and fifth chakras. Perhaps this is where the confusion in esoteric teachings stems from. What appears to happen during incarnation is that these bodies, the emotional and mental, cross. Gifted psychic Sheradon Bryce, in *Joyriding the Universe*, speaks of emotions as degraded forms of feelings. Thus the pure feelings associated with the supramental emotional body become, during incarnation, the mixed emotions experienced on this side of the "veil of forgetting."

Q: Why does Potentiation take nine months to unfold?

A: It takes just over nine months for Potentiation to repattern the bioenergy blueprint and fill it with torsion Source energy because this process is keyed to the physical density of the body and is subject to its timeline. The body is wise and knows exactly what to do when its DNA is functioning harmonically. We like to think of Potentiation Electromagnetic Repatterning as a "rebirth cycle," making the nine-month (42-week) period most appropriate.

Q: Does the Regenetics Method take care of Candida?

A: A popular misconception even in the alternative health community is that Candida is "bad." Nothing could be further from the truth. *Candida albicans* is one of many important microorganisms in the body that are "saprophytic," meaning it consumes dead and toxic tissue, and when it proliferates it actually is trying to cleanse not harm the body. Candida overgrowth problems are often reported to lessen as the body detoxifies over the course of Regenetics. This makes perfect sense because as toxicity levels decline, there is less reason for Candida to proliferate systemically.

Q: What do you see as the relationship between the electromagnetic fields and chakras?

A: Chakras are bioenergy centers in the form of wheels running vertically along the spinal column and head. The chakra system is activated progressively, leading to increased bioenergy output and availability, in the Regenetics Method. Each of the principal chakras corresponds numerically to an electromagnetic field, and functions in tandem with that field, such that the first or "root" chakra aligns with the first electromagnetic field, the second or "sex" chakra aligns with the second field, etc. Together, the chakras (which process hyperdimensional torsion energy in the form of light) and electromagnetic fields (which process hyperdimensional torsion energy in the form of sound) establish the holographic interface that gives rise to the human body. Utilizing a genetic sound-light translation mechanism detailed in Chapter Six, each sonic field energizes the corresponding chakra, which then transfers bioenergy to specific aspects of the subtle anatomy. The majority of elements governed by a particular electromagnetic field also apply to the corresponding chakra.

Q: Does everyone benefit from the Regenetics Method?

A: We have had the honor and pleasure to work with people from a variety of backgrounds and beliefs. Some have had more profound results than others. It seems that anyone approaching this work with a truly open mind and especially heart experiences a positive shift—even if it was not what was expected. We advise that you set your intention specifically yet flexibly on what you want to achieve, then trust the wisdom of your DNA as you go about living your life with joy. Good ways to maintain intention include regular meditating, praying and journaling on your own conscious personal mastery as it expresses itself in your individualized evolutionary path. Your mind is extremely powerful, so use it to create a healthier reality. Finally, we cannot overemphasize the importance of inviting more and more love into all levels of your being, as "exercising the heart" ultimately is what makes your DNA available for activation and transformation.

Q: Can someone who is already healthy benefit from Regenetics?

A: How do you define "healthy"? For us, *healthy* describes someone who is whole in every way: physically, mentally, emotionally, and spiritually. Sadly, from this perspective, few people today are healthy. Clients with no physical problems often report substantial healing on mental, emotional and/or spiritual levels. Others experience the Regenetics Method more palpably. No two individuals are alike, but most clients (even those who consider themselves healthy) report positive shifts in one or more areas.

Q: I've been reading about the healing power of *kundalini* and am interested in Articulation Bioenergy Enhancement. Is it possible to receive Articulation without having done Potentiation?

A: Typically, no. On very rare occasions, we have worked with individuals who already have succeeded in sealing their Fragmentary Body, making it possible to skip Potentiation and move directly to Articulation. Generally speaking, however, Potentiation is the primary DNA activation that makes Articulation (as well as Elucidation and Transcension) possible. One needs to be at the five-month mark of Potentiation or beyond in order to benefit from Articulation. Elucidation Triune Activation is then appropriate after the 42-week Potentiation cycle has completed, and Transcension Bioenergy Crystallization can be experienced after the 42-week Elucidation cycle finishes. Articulation, Elucidation and Transcension can be performed later than this minimum timeline—without diminishing their effectiveness—but not earlier.

Q: How do I know if the Regenetics Method is right for me?

A: Trust your intuition. We live in a world based largely on denying our own power. The Regenetics Method involves experiencing firsthand the transformational truth that the only real power exists in and through you. If you are afraid to change; feel locked in victim consciousness; believe that only someone or something outside yourself can "cure" you; or are addicted to old illness or relational patterns, Regenetics probably is not for you—at this time. If, however, the concept of conscious personal mastery excites you; you are committed to your own empowerment; you believe it is possible to transcend limitation; and the Regenetics Method resonates with you—go for it!

Q: Do I need to commit to the full Regenetics Method in order to maximize my benefits?
A: Although many of our clients have benefited greatly from only Potentiation or only Potentiation combined with Articulation, we strongly recommend, at a minimum, that you consider committing to the Core Regenetics Series of Potentiation, Articulation and Elucidation as a path of conscious personal mastery. Dedicated spiritual seekers might consider our Advanced Mastery Program which includes Transcension. In addition to saving the client money, such a commitment represents a more powerfully energized intention that appears to have a more profound impact on the individual's DNA activation and unfoldment. That said, there is certainly nothing wrong with trying Potentiation first before determining whether to move forward.

Q: Do you offer a sliding scale or other discounts for your services?
A: We typically do not discount our fees as we feel they are more than fair for the unique services we provide. We do offer special discounted fees for individuals who commit to the Core Regenetics Series or Advanced Mastery Program. In addition, we perform sessions free of charge for children under twelve as long as at least one parent or guardian is willing to experience the same session by paying our normal fee. Children under twelve

also are "grandfathered in" for the applicable future sessions when either the Core Regenetics Series or Advanced Mastery Program is purchased. Parents have reported significant benefits even in very young children, who by nature engage the Regenetics Method less cognitively and more intuitively than adults.

Q: Do you ever perform sessions for groups?
A: We often work with couples and even whole families. In November, 2008, we offered a complimentary planetary session called Global Potentiation Day, which simultaneously reached thousands of people around the world and inspired a great deal of inspirational feedback. Doing a collective session not only is capable of facilitating individual issues, but also of working morphogenetically to heal one-on-one and group relational dynamics. This is an extremely exciting application of Regenetics, tremendously broadening the scope of this Method by showing it can used not just on an individual basis, but equally for "couples therapy," "family counseling," and even "community building."

Q: Do you ever teach others how to perform Regenetics?
A: For existing clients, we offer regular Regenetics Seminars where you can learn to facilitate the four primary phases of the Regenetics Method. For details visit **http://www.phoenixregenetics.org**.

Q: How do I get started?
A: Email (preferable) or telephone us using the updated contact information provided on either of our two main websites: **http://www.phoenixregenetics.org** or **www.potentiation.net**. We will need your name, date of birth, location (time zone), and email address. Happy potentiating!

APPENDIX C: SAMPLE ELECTROMAGNETIC GROUP

This Appendix provides a simplified Schematic of the first Electromagnetic Group encountered by the developers of the Regenetics Method. There are twelve such groups in total corresponding to the twelve pairs of cranial nerves, suggesting that the twelve groups together compose the collective Mind of humanity. For various reasons, *a large percentage of those attracted to Regenetics exhibit the "ener-genetic" structure sketched in this sample Schematic.*

Following Potentiation Electromagnetic Repatterning, during which surrogate muscle testing is employed to determine the correct energy family, clients receive a similarly detailed Schematic of the particular Electromagnetic Group to which they belong.

This Schematic, which is given for educational purposes to provide a "roadmap" for the cyclical movement of "torsion" energy through the bioenergy centers following Potentiation, is also a helpful reference for Articulation Bioenergy Enhancement, Elucidation Triune Activation, and Transcension Bioenergy Crystallization.

Not counting the Master or Source Field, which corresponds to our ultimate God-beingness in Galactic Center, the nine electromagnetic fields outlined in this Schematic represent a gestalt or "ecosystem" where a number of elements interrelate either harmoniously to produce vitality or disharmoniously to create disease. In the latter case, dysfunction in a particular field can, but *does not necessarily,* result in one of several Focal Conditions.

IMPORTANT: These Focal Conditions represent *generalized potentials only* for the Electromagnetic Group and may have nothing whatsoever to do with an individual's situation. In other words, while these Focal Conditions indicate the locations of bioenergy disruptions in the electromagnetic fields that give rise to common medical conditions, **Focal Conditions do not constitute individualized diagnosis**. In no case is the information provided in this Schematic intended to diagnose any medical condition or recommend any medical treatment, medication, or supplement.

There is nothing the client is expected to "do" with this Schematic, which primarily is intended to help "ground" Potentiation and the Regenetics Method with specific data and a time frame relative to each electromagnetic field. For example, in the event of emotional release, by referencing the Schematic, it may be possible to determine whether the emotions in question relate to the electromagnetic field currently being energized by torsion waves (universal creative consciousness).

Included are concepts from many disciplines ranging from allopathic medicine and homeopathy to Chinese medicine and astrology that function within the electromagnetic blueprint and may or may not assist with understanding—depending on the individual's orientation and background. *It is unnecessary to understand all the terms contained in this Schematic, which is designed purposely to meet you at your level.*

The Regenetics Method recognizes that the human bioenergy fields are primary in creating health or disease, an idea substantiated by recent research in "wave-genetics" proving that cellular functions are regulated not just biochemically, but bioenergetically. Therefore, it is helpful to "map" these fields in order to understand them individually and in holistic relationship to one another. This is an empowering approach to grasping the complexity of the human organism across the body-mind-spirit continuum, as opposed to seeing ourselves as limited physical beings.

The order presented in this Schematic charts the flow of torsion energy from the Master Field through the nine electromagnetic fields (and corresponding *chakras*) which Potentiation initiates. Starting with the ninth field, "potentiators" spend an average of ten days in each field on the way "down" (9-1), then approximately seven days per field on the way back "up" (meaning a total of seventeen days in the first field as the torsion waves reverse direction).

Once the ninth field is reached again, there follows a transitional period of eleven days or so as the number of fields recalibrates from nine to eight in number. During this period, the ninth field literally descends and fuses with the second field, creating an "infinity circuit" that makes possible the roughly four-month "charging" phase when each field from the eighth down slowly fills with Source-derived torsion energy much like a tiered fountain as shown in Figure 4. At this point, *approximately five months from the date of Potentiation, one becomes eligible to experience Articulation.*

The tenth field, labeled the Master (Source) Field, is the unified field of the soul that is not subject to fragmentation or disease and is and always has been "whole." In a profound sense, Potentiation is designed as the first step in assisting individuals to unite this Source Field with their lower fields, a transformational process that continues with Articulation, Elucidation, and Transcension. The ultimate goal is to follow this path of conscious personal mastery to complete bio-spiritual healing and enlightenment.

Blanks (—) in the Schematic indicate no applicable energy for the category.

FIELD: Master (Source)
CHAKRA: —
RAY (TRUE COLOR): —
PERMANENT ATOM: Atmic
SUBTLE BODY: Soul
GLAND: —
BRAINWAVE: —
GENETIC/CELLULAR ASPECT: —
ORGAN SYSTEM: Nadis
MERIDIAN: —
MIASM: —
PRIMARY TOXIN: —
MICROORGANISM POPULATION: —

EMOTIONS: Gratitude, Faith, Joy, Love
FOCAL CONDITION: —
NUTRIENT: —
CELL SALTS: Calcium Fluoride, Kali Phosphate
TREE OF LIFE: Nezah/Eternity
ACTIVE PRAYER MODALITY: —
PLANETS: Mars, Venus
SIGNS: Aries, Libra

ELECTROMAGNETIC FIELD: 9
CHAKRA: 9
RAY (TRUE COLOR): —
PERMANENT ATOM: —
SUBTLE BODY: Spiritual
GLANDS: Salivary
BRAINWAVE: —
GENETIC/CELLULAR ASPECT: DNA
ORGAN SYSTEMS: Autonomic Nervous System, Gall Bladder, Liver
MERIDIAN: Liver/Gall Bladder
MIASM: —
PRIMARY TOXIN: —
MICROORGANISM POPULATION: —
EMOTIONS: Atonement, Deprivation, Resentment, Sense of Being Trapped, Unforgivingness
FOCAL CONDITIONS: Anemia, Liver Disease, Multiple Sclerosis (MS), Neuralgia, Neurosis, Parkinson Disease
NUTRIENTS: Hydrogen, Nitrogen, Sulfur
CELL SALT: Natrium Mur
TREE OF LIFE: Yesod/Foundation
ACTIVE PRAYER MODALITY: —
PLANET: Sun
SIGNS: Leo, Ophiuchus

ELECTROMAGNETIC FIELD: 8
CHAKRA: 8
RAY (TRUE COLOR): White
PERMANENT ATOM: —
SUBTLE BODY: Spiritual (Lightbody)
GLANDS: Hypothalamus, Lacrimal
BRAINWAVE: —
GENETIC/CELLULAR ASPECT: Mitochondrial DNA
ORGAN SYSTEMS: Sinus/Limbic, Olfactory
MERIDIANS: Central Vessel, Governing Vessel
MIASM: —

PRIMARY TOXIN: —
MICROORGANISM POPULATION: —
EMOTIONS: Despair, Grief, Guilt, Melancholy, Yearning
FOCAL CONDITIONS: Depression, Sinusitis, Seasonal Affective Disorder (SAD), Snoring
NUTRIENTS: Potassium, Salt, Sodium, Trace Minerals
CELL SALT: Kali Mur
TREE OF LIFE: Keter/Crown, Hokhmah/Wisdom
ACTIVE PRAYER MODALITY: —
PLANETS: Earth, Moon
SIGNS: Capricornus, Scorpius

ELECTROMAGNETIC FIELD: 7
CHAKRA: 7
RAY (TRUE COLOR): Violet
PERMANENT ATOM: Mental
SUBTLE BODY: Mental
GLAND: Parathyroid
BRAINWAVE: Gamma
GENETIC/CELLULAR ASPECTS: Cytosine, RNA, Cell Division, Krebs Cycle
ORGAN SYSTEMS: Bladder/Kidney/Urinary, Musculoskeletal
MERIDIAN: Bladder/Kidney
MIASMS: Vaccination, Will
PRIMARY TOXINS: Antibiotics, Chem Trails, Fluoride, Root Canal Toxins, Vaccines
MICROORGANISM POPULATION: Intestinal Flora (Candida)
EMOTIONS: Apathy, Disappointment, Discouragement, Disillusionment, Frustration, Helplessness, Hopelessness, Lack of Faith, Stress
FOCAL CONDITIONS: ADD/ADHD, AIDS, Arthritis, Autism, Avian Flu, Cavitation, CIFDS (CFS), Fibromyalgia, Gulf War Syndrome, Incontinence, Leukemia, Lupus, Multiple Chemical Sensitivity (MCS), Osteoporosis, Scoliosis, SARS, SIDS
NUTRIENTS: Activator X, Boron, Calcium, Ionic Calcium, Ferrous Sulphate, Iron, Magnesium, Vitamin D, Vitamin D3
CELL SALTS: Magnesium Phosphate, Natrium Sulphate
TREE OF LIFE: Hesed/Mercy
ACTIVE PRAYER MODALITY: Thought
PLANETS: Saturn, Vulcan
SIGN: Pisces

ELECTROMAGNETIC FIELD: 6
CHAKRA: 6
RAY (TRUE COLOR): Indigo

PERMANENT ATOM: Astral
SUBTLE BODY: Mental
GLANDS: Sweat
BRAINWAVE: Theta
GENETIC/CELLULAR ASPECT: Adenine
ORGAN SYSTEMS: Auditory, Dermal, Mucous Membrane, Respiratory
MERIDIAN: Lung
MIASMS: Psora, Tuberculosis
PRIMARY TOXINS: Airborne Allergens, Bacterial Toxins, Heavy Metals, Metallic Dental Materials
MICROORGANISM POPULATION: Bacteria, Mycobacteria, Mycoplasmas, Spiroplasmas
EMOTIONS: —
FOCAL CONDITIONS: Acne, Asthma, Bronchitis, Eczema, Inner Ear Infection, Moles, Psychosis, Psoriasis, Rashes, Suffocation, Tinnitus, Vertigo
NUTRIENTS: Iodine, Molybdenum, Vitamin B, Vitamin B_{12}
CELL SALT: Calcium Sulphate
TREE OF LIFE: Tiferet/Beauty
ACTIVE PRAYER MODALITY: Love
PLANET: Mercury
SIGN: Gemini

ELECTROMAGNETIC FIELD: 5
CHAKRA: 5
RAY (TRUE COLOR): Blue
PERMANENT ATOM: Buddhic/Christ
SUBTLE BODY: Emotional
GLAND: Pituitary
BRAINWAVE: Alpha
GENETIC /CELLULAR ASPECT: Thymine
ORGAN SYSTEM: Circulatory
MERIDIANS: Heart, Pericardium
MIASMS: Syphilitic, Thuja Focal
PRIMARY TOXINS: Hydrocarbons, Chlorinated Hydrocarbons
MICROORGANISM POPULATION: Homeostatic Soil Organisms (HSOs)
EMOTIONS: Ambition, Desire, Lust
FOCAL CONDITIONS: Arteriosclerosis, Endocrine Imbalances, Heart Disease, Hemophilia, Hot Flashes, Hypertension, Wilson Disease
NUTRIENTS: Manganese, Selenium, Vitamin E, Zinc
CELL SALT: Silicea
TREE OF LIFE: Malkhut/Kingdom

ACTIVE PRAYER MODALITY: Emotion
PLANET: Chiron
SIGN: Sagittarius

ELECTROMAGNETIC FIELD: 4
CHAKRA: 4
RAY (TRUE COLOR): Green
PERMANENT ATOM: —
SUBTLE BODY: Emotional
GLAND: Pineal
BRAINWAVE: Beta
GENETIC/CELLULAR ASPECT: Guanine
ORGAN SYSTEMS: Brain, Central Nervous, Optical
MERIDIAN: Triple Heater
MIASMS: Gonorrhea, Psychotic
PRIMARY TOXINS: Artificial Sweeteners, Cooked Food Toxins, Food Additives, Food Colorings, Genetically Modified Organisms (GMOs), Processed Sugars
MICROORGANISM POPULATION: Yeasts
EMOTIONS: Abandonment, Arrogance, Betrayal, Confusion, Pride, Rejection
FOCAL CONDITIONS: ALS, Alzheimer Disease, Cataracts, Diabetes, Dyslexia, Encephalitis, Food Allergies, Glaucoma, Hypoglycemia, Insomnia, Migraine, Obsessive-compulsive Disorder (OCD)
NUTRIENTS: Chromium, Methionine, Vanadium, Vitamin K
CELL SALT: Kali Sulphate
TREE OF LIFE: Daat/Knowledge
ACTIVE PRAYER MODALITY: Feeling
PLANET: Jupiter
SIGN: Aquarius

ELECTROMAGNETIC FIELD: 3
CHAKRAS: 3, Pranic Triangle
RAY (TRUE COLOR): Yellow
PERMANENT ATOM: Physical
SUBTLE BODY: Physical
GLANDS: Adrenal, Thymus
BRAINWAVE: Delta
GENETIC/CELLULAR ASPECT: Uracil
ORGAN SYSTEM: Immune
MERIDIAN: Spleen
MIASMS: Cancer, Radiation

PRIMARY TOXINS: Chemicals, Mechanized Fields, Pharmaceuticals, Radioactive Metals, Recreational Drugs, Smoke, Solvents
MICROORGANISM POPULATION: Viruses
EMOTIONS: Anxiety, Fear, Lack of Trust, Panic, Terror, Worry
FOCAL CONDITIONS: Cancer, Lowered Immunity, Paranoia
NUTRIENT: Vitamin C
CELL SALT: Calcium Phosphate
TREE OF LIFE: Hod/Reverberation
ACTIVE PRAYER MODALITY: Peace
PLANET: Pluto
SIGN: Virgo

ELECTROMAGNETIC FIELD: 2 (the Fragmentary Body)
CHAKRA: 2
RAY (TRUE COLOR): Orange
PERMANENT ATOM: —
SUBTLE BODY: —
GLAND: Thyroid
BRAINWAVE: —
GENETIC/CELLULAR ASPECT: —
ORGAN SYSTEMS: Oral, Reproductive
MERIDIAN: —
MIASM: —
PRIMARY TOXINS: Bacterial Toxins, Parasitic Toxins
MICROORGANISM POPULATIONS: Dental Bacteria, Parasites
EMOTIONS: Envy, Jealousy, Shame
FOCAL CONDITIONS: Dental Decay, Fibroids, Halitosis, Impotence, Parasitic Infection, Infertility, Periodontal Disease, Possession, Reproductive System Illness, Sterility
NUTRIENT: —
CELL SALT: Natrium Phosphate
TREE OF LIFE: Gevurah/Judgment
ACTIVE PRAYER MODALITY: —
PLANET: Uranus
SIGN: Taurus

ELECTROMAGNETIC FIELD: 1
CHAKRA: 1
RAY (TRUE COLOR): Red
PERMANENT ATOM: —
SUBTLE BODY: Physical
GLANDS: Parotid
BRAINWAVE: —
GENETIC/CELLULAR ASPECT: —

ORGAN SYSTEMS: Digestive, Pancreatic
MERIDIANS: Large Intestine, Small Intestine, Stomach
MIASM: —
PRIMARY TOXIN: Mycotoxins (from fungal overgrowths)
MICROORGANISM POPULATION: Fungi
EMOTIONS: Anger, Disgust, Hatred, Rage
FOCAL CONDITIONS: Acid Reflux, Bloating, Colitis, Crohn Disease, Excessive Gas, Fungal Infection, Hemorrhoids, Irritable Bowel Syndrome (IBS), Leaky Gut, Poor Digestion
NUTRIENTS: Vitamin A, Vitamin F, Vitamin P
CELL SALT: Ferrum Phosphate
TREE OF LIFE: Binah/Understanding
ACTIVE PRAYER MODALITY: —
PLANET: Neptune
SIGN: Cancer

GLOSSARY OF TERMS

[The following list of definitions is intended for reference to assist with understanding the more technical aspects of this book. This Glossary also may be read in its entirety as a "journey in consciousness" through a series of related ideas. Note that 1) most capitalized words and phrases have their own separate definitions; and 2) many of the terms listed are, for practical purposes, synonyms.]

Active Prayer: phrase used by Gregg Braden in *The Isaiah Effect* to describe a "prayer technology" employed by the Essenes from the time of Christ. Active Prayer is designed to affect, and effect, quantum outcomes by changing the pray-er's picture of reality through focused intention that validates whatever one is praying for as already having happened. The five intercessory modalities used in Active Prayer—thought, feeling, emotion, peace, and love—correspond to the five Nucleotide bases of DNA and RNA. The nucleobases, in turn, align with the five vowels. Thus when we change reality through linguistically expressed intention, we do so via our bodies by activating our divine genetic endowment.

Adam Kadmon: kabalistic name for the fully activated Lightbody based on the tripartite tetrahedron shape and the Merkabah. The Adam Kadmon represents the next evolutionary unfolding of the transformative genetic potential (the Holy Grail) that is innate to the human genome.

Aether: ancient Greek name for the subspace field of Torsion Energy responsible for universal manifestation. Sometimes spelled "ether," this generative, conscious energy flows like time in a sacred geometric spiral of a fractal nature that has been called Phi, the Golden Mean, and the Fibonacci Sequence. Modern scientists are returning to the notion of Aether using such phrases as

Zero Point Energy. As this hyperdimensional light materializes in the creation of form, it appears to become liquefied light or Plasma.

Age of Consciousness: dawning Aquarian "Age of Light" many believe is precipitating the dissolution of old hierarchical structures of control, fear and manipulation, to be followed by the birth of social models based on principles of partnership, servant leadership, Unity Consciousness, and Unconditional Love.

Articulation (Bioenergy Enhancement): second DNA activation in the Regenetics Method. Appropriate as of the five-month mark following Potentiation Electromagnetic Repatterning, after the Fragmentary Body has experienced Sealing, Articulation is designed to stimulate bioenergy or Kundalini at the genetic and cellular levels, enhancing creativity while facilitating the transformation of limiting thought-forms and behaviors related to distortions in the mental Subtle Body.

Atmic Permanent Atom: stable force center that is the soul portion of our being. Collectively, the Permanent Atoms (around which the Subtle Bodies form) serve as data memory banks for establishing the educational circumstances (Karma) of a given incarnation. In addition to the Atmic, the Permanent Atoms include the spiritual, emotional, mental and physical, which correspond to the four interrelated Subtle Bodies that are activated progressively in the Regenetics Method. While the spiritual, emotional, mental and physical Subtle Bodies align with particular lower Electromagnetic Fields in dualistic pairs, the Atmic Permanent Atom remains unified in the Master or Source Field as the divine aspect of ourselves that never experiences separation and loss of Unity Consciousness through the "veiling" or forgetting process associated with Duality and the Fragmentary Body.

Atonement: often misunderstood term that can be read as "at-one-ment," or Healing into wholeness

(Enlightenment) by returning bio-spiritually to Unity Consciousness.

Aura: prismatic halo of bioenergy surrounding the human body composed of interconnected auric or Electromagnetic Fields. These fields provide an index of the relative health of the organism's underlying energetic structure. The Aura represents both a bioenergetic and Multidimensional blueprint.

Autoimmunity: degenerative condition often induced through genetic alteration (reverse Transcription) by invasive factors such as vaccines and genetically modified foods (GMOs) in which the immune system begins attacking the body's own toxic cells.

Biological Terrain: internal measure of microorganism levels in relation to one another. Biological Terrain, said to be balanced or imbalanced, is determined by such factors as stress, toxicity, antibiotics use, and vaccine history.

Biophoton Light Communication: data communication network used by living organisms essential to proper biological functioning and immune response that employs light to transmit and receive information. The cellular Hologram equivalent of the nervous system, the system of Biophoton Light Communication operates far more quickly than the nervous system and may be thought of as a parallel-processing "biocomputer" allowing for an unmediated energetic interface with the individual's environment.

Black Hole: region of space-time said to result from a collapsed supernova from which it once was thought not even light could escape. Recently, however, physicist Stephen Hawking admitted that Black Holes may leak information. Many theorize that as the Black Hole at Galactic Center becomes more energized, it catalyzes human evolution through Torsion Energy emissions transmitted via the Photon Band by way of the sun—an idea consistent with the ancient notion that Galactic Center is the womb or Source of life. See also White Hole.

Black Road: ancient Mesoamerican name for the Multi- and ultimately Transdimensional "road" leading back to our true "home" in Source. The Black Road can be conceptualized as simultaneously an astrophysical and Meta-genetic alignment with Galactic Center.

Chakra: bioenergy locus designed to process hyperdimensional light (Torsion Energy) in the form of a wheel connected to the Subtle Bodies of humans. The Regenetics Method recognizes nine principal Chakras. Each of these corresponds numerically to an Electromagnetic Field and functions in tandem with it, such that the first or "root" Chakra aligns with the first field, the second or "sex" Chakra aligns with the second field, etc. The majority of elements governed by a particular Electromagnetic Field also apply to the corresponding Chakra. Together, the Chakras and Electromagnetic Fields establish the holographic interface that gives rise to the human form.

Child of Light: ancient Mesoamerican phrase for the next evolutionary stage of humanity set to occur in conjunction with the close of the Mayan calendar around 2012. The Child of Light or Adam Kadmon inhabits a fully activated fourth-dimensional Lightbody and acts out of Unity Consciousness and Unconditional Love.

Clearing: term for energetic allergy treatment popularized by Dr. Devi Nambudripad, developer of NAET. The term Clearing also has been applied to emotional release work. Sometimes used as a synonym for DNA Activation in the Regenetics Method.

Conscious Personal Mastery: Unconditional Love of oneself as simultaneously the Creator and the created extended outward to all perceived others. The Regenetics Method is designed specifically to promote Conscious

Personal Mastery as a bio-spiritual Healing path leading to Enlightenment.

Creation Beam: see Photon Band.

Curing: act of alleviating a symptom or ailment in another person or in oneself without necessarily assisting the sufferer to achieve Healing.

Cymatics: study of the effect of sound waves on physical (including molecular) form.

Dark Matter: phrase recently coined by scientists to indicate the invisible ninety percent or more of the universe, which apparently resides in other dimensions. It has been theorized that Dark Matter, also referred to as the "space matrix" and "quantum potential," is Torsion Energy. Dark Matter is a way of conceptualizing the hyperdimensional light (often called "black light" in shamanic traditions) that becomes Plasma during physical manifestation.

Descension: alternative way to conceptualize ascension emphasizing that evolutionary transfiguration, rather than an earthly exit, can be a process of grounding higher-dimensional consciousness in physical form by creating the Lightbody.

DNA: deoxyribonucleic acid. Technically a salt (sodium) and thus an excellent conductor of electromagnetism, DNA is located primarily in the mitochondria and nuclei of plant and animal cells, forming the genetic code in sixty-four three-letter combinations of Nucleotides called codons.

DNA Activation: electrogenetic mode of intercession capable of noninvasively stimulating a self-healing potential in the genome, specifically by stimulating rearrangement of Transposons in Potential DNA. The Regenetics Method employs four integrated DNA Activations. Also called Wave-genetics.

DNA Phantom Effect: discovery made famous by the Gariaev group in Russia that wave-activated DNA is capable of communicating outside space-time by creating wormholes as channels for transmission and reception of Universal Creative Consciousness or Torsion Energy.

Duality: illusory divided state of being that has abandoned Unity Consciousness in favor of binary thinking and fragmentation. Duality also refers to the holographic system (so-called reality as most humans currently experience it) birthed by this divided consciousness. In humans Duality imprints and sustains itself in the Fragmentary Body.

Electrogenetics: see DNA Activation.

Electromagnetic Field: one of a set of high-frequency bands of Torsion Energy composed of sound in hyperdimensional octaves, referred to collectively as the Aura and matching the Chakras in order and number. Each Electromagnetic Field, in tandem with the corresponding Chakra, governs a set of related functions in humans. Prior to Potentiation Electromagnetic Repatterning, most people have an unstable structure of nine Electromagnetic Fields (and Chakras) that initiate at the physical level and become increasingly "subtle." During Potentiation the Electromagnetic Fields and Chakras recalibrate from nine to an Infinity Circuit based on the alchemically transformative number 8 as the Fragmentary Body is transformed through Sealing.

Electromagnetic Group: phrase coined by the developers of the Regenetics Method to indicate any of twelve family groups of humans sharing a unique bioenergy structure at the level of the Electromagnetic Fields and corresponding Chakras.

Elucidation (Triune Activation): third DNA Activation in the Regenetics Method focused on the emotional Subtle Body designed to activate a mostly dormant portion of the prefrontal lobes by way of the neocortex and triune brain, establishing the "ener-genetic" precondition—the Unified Consciousness Field—for Lightbody creation. Elucidation, appropriate following Articulation Bioenergy Enhancement as of the 42-week mark of Potentiation Electromagnetic Repatterning, encourages Conscious Personal Mastery by assisting the individual to replace limiting and/or harmful emotions and beliefs with life-affirming ones.

Energized Narrative: phrase coined by the developers of the Regenetics Method to describe a particular combination of sound and intention (in the form of special vowel-based words) delivered in the form of hyperdimensional Spiral Standing Waves to initiate DNA Activation.

Enlightenment: not to be confused with the so-called enlightenment of many Eastern mindfulness practices, here Enlightenment is defined as the process and result of allowing in higher-dimensional light to the point that one literally becomes a higher-dimensional being. Genuine Enlightenment follows a path of Conscious Personal Mastery and results from Healing or "wholing" through the actual, physical embodiment of Unity Consciousness. By definition, Enlightenment involves creating a Lightbody.

Epigenetic: adjective popularized by biologist Bruce Lipton that signals an important development in the field of genetics from a focus on the cell nucleus and "nature" as the sole evolutionary forces worth studying, to the cell membrane and "nurture" as more primary instigators of evolution. Readers of *Conscious Healing* will note that the author considers the Meta-genetic an even more fundamental basis for the evolution of species than either the genetic or Epigenetic.

Exon: coding segment of genes active during biochemical genetic replication involving RNA Transcription of DNA codes.

Fibonacci Sequence: see Golden Mean.

Fragmentary Body: name given to the second Electromagnetic Field and corresponding Chakra in humans. The Fragmentary Body is a "Frankenstein's monster" of energies, many of which do not belong in the body. The energy for parasites, for example, attaches to the second Electromagnetic Field and Chakra. The Fragmentary Body is an anti-Enlightenment "rift" in the human bioenergy fields that limits one's ability to embody Unity Consciousness by activating the Lightbody. During Potentiation Electromagnetic Repatterning, the Fragmentary Body is transformed through Sealing, which represents a critical first step in bridging the perception of Duality at the biological level.

Frequency Domain: phrase coined by physicist David Bohm equivalent to dimension that indicates a certain frequency range of the electromagnetic spectrum. With each of the Electromagnetic Fields occupying a particular Frequency Domain, the Aura can be said to contain the human Multidimensional blueprint. The Regenetics Method recognizes eight perceptible Frequency Domains that can be made to align through Potentiation Electromagnetic Repatterning with the numerically corresponding Electromagnetic Fields and Chakras.

Galactic Center: astrophysical name for the Creator, the Transdimensional Source or "engine" of linguistically based, Meta-genetic evolution in the Multidimensional universe, located between the Black Hole and White Hole at the core of the Milky Way Galaxy. Also called Healing Sun, Logos, and Tula.

Galactic Year: one of several cycles of time ending with the close of the Mayan calendar around 2012. It takes roughly 225 million Earth years for our galaxy to make one complete rotation through the Photon Band, which is believed to be a galactic "birth cycle." At the beginning of this cycle, Earth's landmass, Pangaea, began separating into the seven continents. This process of planetary Individuation correlates with continental drift theory.

Ge: name for the omnipotent creational "tone" of Unconditional Love, the primary, Meta-genetic "language" of Torsion Energy emanating from the Silent Stillness of Galactic Center. This divine "frequency," related to the biblical Word, can both create life and promote genuine Healing and Enlightenment by activating the Holy Grail (Lightbody) potential that lies dormant in Potential DNA. The name "Ge" most likely stems from the G-shape of the Milky Way Galaxy as it spirals outward from Source, during which time Ge differentiates into Spiral Standing Waves of hyperdimensional sound and light—in that order.

Genetic Sound-light Translation Mechanism: phrase coined by the developers of the Regenetics Method to indicate the process by which chromosomes assemble themselves into a solitonic lattice designed to "translate" highly stable waves of sound (phonons) into light (photons), and vice versa. The conception of the human body as a Hologram depends on such a mechanism, a related function of which also allows hyperdimensional sound to become light during physical manifestation.

Golden Mean: Phi ratio of 1.6180339... that serves as a cornerstone for sacred geometry, from the mathematics of the spiraling DNA molecule to the spiraling structure of the galaxy. Also known as Fibonacci Sequence.

Green Language: musical, vowel-only language of Meta-genetic transformation. Medieval alchemists used this phrase probably owing to the alteration of hydrogen bonds, which are green on the electromagnetic spectrum, in biological water molecules that occurs during fourth-dimensional Lightbody creation. Also called *lingua adamica* and Language of the Birds.

Healing: ultimately self-directed process following a path of Conscious Personal Mastery and leading to Enlightenment that involves transcending the perception of Duality by returning to Unity Consciousness, reuniting with Source as an individuated being, and becoming "whole."

Healing Sun: ancient Mesoamerican name for Galactic Center that highlights the restorative quality of the Torsion Energy manifesting as hyperdimensional light emitted by the Central Sun of Source.

Heisenberg Uncertainty Principle: discovery popularized in Quantum physics that a scientist always and inevitably affects—and even may effect—the outcome of an experiment simply by observing it.

High-spin Metal: one of six precious metals including gold, iridium, osmium, palladium, platinum and rhodium forming part of the human brain and central nervous system at the Multidimensional level. This unique family of metals can be made to align their atomic spins by the six notes of the Solfeggio Scale combined with the Language of the Birds to stimulate an internal source of Torsion Energy called Kundalini that manifests as hyperdimensional light. It is theorized that the High-spin Metals are stimulated monatomically in Elucidation Triune Activation to establish the Unified Consciousness Field, considered the ener-genetic precondition for Lightbody creation.

Hologram: illusion of form projected by intersecting electromagnetic waves. In a Hologram, the part always contains the whole (the modern way of saying "As above, so below"), everything is energy, and matter per se does not exist.

Holographic Model: phrase based on the theories of Richard Alan Miller, Karl Pribram, David Bohm and others that evokes the Hologram-like nature of reality. A holographic universe is composed of intersecting waves of sound and light that combine to project the illusion of matter.

Holy Grail: human genetic potential for Enlightenment or metamorphosis into a fourth-dimensional Lightbody that emerges when Potential DNA is keyed to create a crystalline, Merkabah-based "Song Grail" or "love song in the blood" in tune with Galactic Center's signature tone of Ge.

Hypercommunication: extrasensory communication similar to telepathy that transcends spatial and temporal limitations used, for example, in ant colonies. Also available to humans, Hypercommunication operates via the "biological Internet" of DNA. The Regenetics Method can be thought of as a practical application of Hypercommunication.

Individuation: process and result of separating from Source and achieving individual consciousness. After people individuate, it becomes possible for them to return to Unity Consciousness, and ultimately Source, as fully realized individuals who have achieved Healing and Enlightenment.

Infinity Circuit: ener-genetic blueprint of our authentic divine nature based in Unity Consciousness patterned on the alchemically transformative number 8. During Potentiation Electromagnetic Repatterning, the Electromagnetic Fields and Chakras recalibrate from nine to eight in number as the Fragmentary Body is transformed through Sealing. The Infinity Circuit then can be married to the Unified Consciousness Field, establishment of which is the precondition for evolutionary transmutation into the Lightbody.

Intron: noncoding segment of genes and primary aspect (with Transposons) of Potential DNA. When Transcription enzymes finish transcribing a gene, editing enzymes remove the Introns and splice together coding segments called Exons.

Karma: law and operating principle of the holographic multiverse designed to teach individuals responsibility for the full range of their creations.

Key of Life: device also known as an ankh featured in Egyptian hieroglyphics. The Key of Life may have been a type of actual tuning fork for harmonizing with Galactic Center's signature tone of Ge, or may have symbolized techniques (such as use of the Solfeggio Scale) for producing this celestial harmonization. In either case, the Key of Life is of a musical nature and designed to be employed with the Language of the Birds to stimulate Potential DNA to create the Lightbody.

Kinesiology: science of muscle testing. Kinesiology employs muscle-response (strong or weak) tests to determine allergies, emotional blockages, and even the truth or falsehood of given statements. Since its invention in the 1960s, Kinesiology has become popular among both alternative and mainstream healthcare professionals.

Kundalini: human "life-wave" of Torsion Energy that lies mostly dormant technically just below the second Electromagnetic Field and corresponding Chakra until activated. Before Sealing occurs, the Fragmentary Body acts as a "glass ceiling" inhibiting the upward flow of Kundalini that occurs during the process of Healing. In Vedic teachings, Kundalini is considered the highest evolutionary force capable of unfolding one's full bio-spiritual potential of Enlightenment (the Lightbody) when awakened. Following Sealing of the Fragmentary Body through Potentiation Electromagnetic Repatterning, Articulation Bioenergy Enhancement is designed to stimulate and integrate Kundalini gently, starting at the genetic and cellular levels.

Language of the Birds: vowel-only language of a Meta-genetic and musical nature (see Solfeggio Scale) historically employed by master geneticists such as Jesus to activate Potential DNA to create the Holy Grail or

Lightbody. Also called *lingua adamica* and Green Language.

Lightbody: higher-dimensional physiology also known as the Adam Kadmon and Child of Light structured on the tripartite tetrahedron shape and the Merkabah, designed to hold the light of Unity Consciousness and Unconditional Love. The concept of Lightbody is generalized somewhat in this book. In reality, the biochemical vehicle for consciousness appropriate to each dimension above our own is a form of Lightbody. In other words, we inhabit a series of increasingly "conscious" or light-filled Lightbodies on our evolutionary journey back to complete, undifferentiated unity with Source. For the most part, when the Lightbody is discussed, the fourth-dimensional version that will be activating for the majority of those "harvestable" around 2012 is intended.

Logos: see Galactic Center.

LOVEvolution: term coined by Barry and Janae Weinhold to indicate that the driving force behind evolution is the primary Torsion Energy of Unconditional Love.

Master Field: name given by the developers of the Regenetics Method to the "field" where the chief of the five Subtle Bodies, the Soul Body, resides. The Master Field, centered on the Atmic Permanent Atom, exists outside the Multidimensional (holographic) electromagnetic and torsion spectrum in a Transdimensional state of Silent Stillness associated with the tone of Ge. Also called Source Field.

Meridian System: Eastern system of subtle energy lines in the human body that serves as a basis for acupuncture and acupressure, which typically recognize twelve principal Meridians.

Merkabah: "trinitized" vehicle of light based on interlocking tetrahedron shapes resembling a three-dimensional Star of David that emerges from within the human form during Lightbody creation.

Meta-genetic: adjective coined by the developers of the Regenetics Method to describe Transdimensional, linguistically based, self-organization functions observed on the microcosmic scale at the level of Introns/Transposons in Potential DNA, and on the macrocosmic level as emanating from Galactic Center and passed on to humans via the Photon Band and the sun. The Meta-genetic is the primary driving force behind evolution of species, far surpassing the reach of both the genetic and the Epigenetic.

Morphogenetic: adjective popularized by biologist Rupert Sheldrake to characterize the ener-genetic field constituted by DNA that connects all biological species regardless of time and distance.

Multidimensional: adjective used to indicate the multi-layered nature of reality beyond what the five senses perceive in three-dimensionality. The human Electromagnetic Fields can be thought of as a geometric matrix that allows access to as many as eight increasingly subtle dimensions or Frequency Domains. This unfolding of perception to the full range represented by the Electromagnetic Fields and corresponding Chakras is what it means to become Multidimensional. See also Transdimensional.

Nadi: one of a network of tiny tubular channels acting like "fiber-optic" extensions of the human nervous system that filter the Torsion Energy of Source by way of the Subtle Bodies into the Electromagnetic Fields and, via DNA, the corresponding Chakras. This bioenergy, directed by the Electromagnetic Fields, passes to specific areas of the subtle anatomy through the Chakras.

Nonlocality: term derived from "nonlocal" employed by physicists to describe Newtonian logic-defying interactions between subatomic particles that take place at a distance.

Nucleotide: one of five combinations of a nucleic acid, a sugar and a phosphoric group that create DNA and RNA used during formation and Transcription of genetic codes.

Ophiuchus: occulted thirteenth astrological sign, also known as the Serpent Bearer and located near Galactic Center, symbolizing DNA's role as a unifying network that links the other twelve astrological signs corresponding to the biblical Twelve Tribes, the twelve pairs of cranial nerves, Earth's twelve tectonic plates, and the twelve Electromagnetic Groups identified by the developers of the Regenetics Method.

Phi: see Golden Mean.

Photon Band: hyperdimensional life-wave of Torsion Energy structured on the Golden Mean connecting Earth to Galactic Center via the sun that serves as a guiding data communication network for human and planetary evolution. Also referred to as Photon Belt and Creation Beam.

Plasma: matter in an electrified state that can be conceptualized as liquefied light, or the form light takes just prior to physical manifestation.

Positronium: atomic particle composed of a negatively charged electron and positively charged positron that illustrates how Lightbody activation occurs by resolving internal Duality. Since electrons and positrons are antiparticle opposites, after combining to form Positronium, they instantly cancel out each other and decay into two particles or Quanta of light.

Potential DNA: phrase coined by the developers of the Regenetics Method to replace "junk" DNA to denote the transformational, Meta-genetic potential that awaits activation in the human genome. Potential DNA interfaces with the life-wave of Torsion Energy emanating from Galactic Center responsible for giving rise to a particular physical form through RNA Transcription of DNA. Evolutionary activation occurs as Potential DNA's Transposons are activated consciously to rewrite or reprogram the genetic code.

Potentiation (Electromagnetic Repatterning): first DNA Activation in the Regenetics Method centered on the physical Subtle Body that initiates an electromagnetic repatterning designed to "reset" the human bioenergy fields at a higher harmonic resonance with Galactic Center. Potentiation also transforms the Fragmentary Body through a process called Sealing, initiating the Healing of the perception of Duality at the biological level.

Quantum: particle of light or other electromagnetic radiation. As an adjective, Quantum refers to the science devoted to the study of subatomic phenomena.

Quantum Potential: phrase coined by physicist David Bohm to refer to the aspect of Nonlocality where space ceases to exist and two electrons, for example, can occupy the same coordinates.

Radionics: instrument- and predominately light-based form of energy medicine that historically has been performed at a distance.

Ray: "true color" associated with a particular dimensional expression of consciousness and physiology in many esoteric traditions. Each Electromagnetic Field and corresponding Chakra corresponds to a native harmonic frequency, or Ray, which manifests as a color of the rainbow. Starting with the first Electromagnetic Field and Chakra and moving up, the Rays are always red, orange, yellow, green, blue, indigo, and violet. Importantly, the eighth "color" in this model is the amalgam of these true colors—white—and corresponds to the eighth or Lightbody Electromagnetic Field and Chakra, responsible for infusing the Unified Consciousness Field with Source energy.

Regenetics (Method): registered service mark for an integrated Method of DNA Activation. As a synonym of Wave-genetics, of which it is a human-potential-based application, Regenetics also describes a field that represents the exciting confluence of energy medicine and molecular biology.

Ribosome: tiny protein structure designed to transfer messenger RNA (mRNA) containing DNA codes out of the cell nucleus during Transcription.

RNA: ribonucleic acid. RNA, in tandem with Ribosomes, is directly responsible for Transcription of DNA codes and protein synthesis.

Samskara: Vedic term describing an ingrained thought-form that maintains one's consciousness in a limited (unenlightened) state. The Fragmentary Body can be thought of as the ultimate Samskara.

Scalar: term coined by Nikola Tesla to indicate thought or intention Torsion Energy capable of traveling faster than observable light. The Regenetics Method theorizes that Scalar waves are identical, for practical purposes, to prana, chi, orgone, Aether, and Kundalini—all of which are forms of hyperdimensional Source energy that, after differentiating from the creational sound current emanating from Galactic Center, manifests as Spiral Standing Waves of hyperdimensional light.

Sealing: term employed by the developers of the Regenetics Method to indicate the stage of ener-genetic repatterning in Potentiation in which the bioenergy vacuum constituted by the Fragmentary Body is closed. Sealing is a critical step on the path to genuine Healing and Enlightenment, as it lays the groundwork for a stable fourth-dimensional Lightbody by establishing an Infinity Circuit of eight Electromagnetic Fields and Chakras.

Silent Stillness: see Unconditional Love.

Solar-planetary Synchronism: phrase coined by Sergey Smelyakov to describe how Earth connects to Galactic Center via our solar system in a harmonic fashion based on Phi or the Golden Mean. See also Photon Band.

Solfeggio Scale: recently rediscovered six-note musical scale believed to contain the exact frequencies used by the Creator to fashion the cosmos in six days. Sacred chants such as the Gregorian once employed these notes to harmonize humanity with Source and increase vitality and longevity. Today's musical scales lack these six frequencies. One of the Solfeggio notes, "Mi," has been used as a frequency by genetic engineers to repair damaged DNA. This is also the note employed in Potentiation Electromagnetic Repatterning.

Somvarta: Vedic term for the powerful, Meta-genetic life-wave of Torsion Energy (Universal Creative Consciousness) emanating from Galactic Center responsible for the spontaneous evolution of species.

Songs of Distinction: for those who have completed the nine-month (42-week) "gestation cycle" of Potentiation Electromagnetic Repatterning and are in need of further assistance with rehydration and remineralization at the level of glands and organs, Songs of Distinction are supplemental ener-genetic fortifications that build on the restored bioenergy blueprint established through the first DNA Activation in the Regenetics Method.

Source: name for the Unified Consciousness Field of Unconditional Love between the Black Hole and White Hole at Galactic Center that uses the omnipotent tone of Ge to create and evolve life in Meta-genetic fashion. Also known as Central Sun, Healing Sun, Logos, and Tula.

Spiral Standing Wave: hyperdimensional Torsion Energy derived from the tone of Ge manifesting as sound and light (in that order) capable of stimulating a Meta-genetic rearrangement of Transposons in Potential DNA from outside space-time, simultaneously promoting Healing and Enlightenment.

Subtle Body: one of five energy bodies in humans denominated physical, mental, emotional, spiritual, and soul. The first four of these sometimes are called the "lesser bodies." The Regenetics Method as a path of Healing leading to Enlightenment through Conscious Personal Mastery progresses through these four bodies— from the physical to the mental to the emotional to the spiritual—in an "archeological" manner designed to access and heal energy distortions rooted in ever deeper and more primary levels of the subtle anatomy.

Superluminal: adjective used to describe anything that moves faster than observable light such as hyperdimensional Spiral Standing Waves of Torsion Energy.

Tachyon: one of a group of Superluminal particle-waves, the discovery of which has challenged many of the assumptions of traditional physics.

Torsion Energy: recently coined scientific term for Universal Creative Consciousness or subspace energy (Aether) experiencing itself in time. In the process of cosmic creation, the primary Torsion Energy of Unconditional Love differentiates into Spiral Standing Waves of hyperdimensional sound and light—in that order—forming a sacred trinity. Torsion Energy in the form of a life-wave (see Photon Band) interfacing with and modifying Potential DNA's Transposons is the Meta-genetic driving force behind the evolution of human consciousness and physiology.

Transcension (Bioenergy Crystallization): fourth and final DNA Activation in the Regenetics Method focused on unblocking and healing distortions in the spiritual Subtle Body, stimulating a safe, progressive

Kundalini awakening to facilitate luminous embodiment of divine consciousness, or Lightbody creation.

Transcription: biological "composition" process in which genetic codes are transferred from one kind of nucleic acid to another, especially from DNA to RNA. In the case of reverse Transcription, which often occurs as a result of vaccination and consuming genetically modified foods (GMOs), RNA rescripts DNA. Transcription is both a biochemically and electromagnetically driven process.

Transdimensional: adjective employed by the developers of the Regenetics Method to describe a unified state of being or Enlightenment, such as that achieved by those in complete harmonic identification with Galactic Center or Source. Transdimensional also may be understood as describing Galactic Center itself, which exists in a state of pure potential or Silent Stillness that transcends, and gives rise to, the Multidimensional holographic universe.

Transposition Burst: phrase coined by biochemist Colm Kelleher to describe a massive molecular rearrangement of Transposons in Potential DNA involving perhaps thousands of genes that occurs during Enlightenment or Lightbody creation.

Transposon: term coined by Nobel laureate Barbara McClintock to describe what also has been called "jumping DNA"—tiny segments of Potential DNA that can be prompted in a Meta-genetic fashion by Torsion Energy or Universal Creative Consciousness to change their positioning in the DNA molecule, rewriting or reprogramming the genetic code.

Triology: term coined by the developers of the Regenetics Method to indicate the evolutionary movement of human biology based on binarisms (Duality) toward a "trinitized" fourth-dimensional consciousness and corresponding physiology (the Lightbody) expressed in the Merkabah by way of the tetrahedron shape.

Tula: ancient Mesoamerican name for Galactic Center or Source believed to be the true home of the god-man Quetzalcoatl. Also referred to as Central Sun, Healing Sun, and Logos.

Unconditional Love: primary Torsion Energy of Source. Theorized in the Regenetics model to be Silent Stillness, the womblike nothingness between the Black Hole and White Hole at Galactic Center, Unconditional Love differentiates into hyperdimensional Spiral Standing Waves of sound and then light (forming a sacred trinity) during manifestation of form. Unconditional Love is the omnipotent, Meta-genetic creational potential driving, by way of the Photon Band, the evolution of human consciousness and physiology.

Unified Consciousness Field: phrase coined by the developers of the Regenetics Method to describe both Galactic Center and the ener-genetic precondition for Lightbody creation. When the individual's Electromagnetic Fields and corresponding Chakras have unified in a gestalt that resonates throughout at Source's signature tone of Ge, manifestation of the Lightbody can occur, theoretically at any time. Elucidation Triune Activation, the third DNA Activation in the Regenetics Method, is designed to assist in the establishment of the Unified Consciousness Field.

Unity Consciousness: defining characteristic of the next evolutionary stage of human consciousness and physiology set to occur collectively in conjunction with the end of the Mayan calendar around 2012. Accessible at any time, Unity Consciousness has been called by many names, including Christ, Buddhic and God consciousness. Unity Consciousness recognizes the divine nature of the self as well as all perceived others and is capable of evolving an enlightened biology in the form of the Lightbody.

Universal Creative Consciousness: term for the creational Torsion Energy of Galactic Center or Source emphasizing that this Meta-genetic energy is, above all else, conscious.

Wave-genetics: term coined by Peter Gariaev to describe his revolutionary research in Meta-genetic medicine. See also DNA Activation and Regenetics.

White Hole: "yang" complement to a "yin" Black Hole theorized to be the procreative aspect of Galactic Center.

Zero Point Energy: phenomenon in which biological organisms use more energy than they can receive from their intake of food, water, and air. This occurs as the distance separating two non-charged surfaces, such as water and a cell membrane, becomes negligible, dimensional coherence takes place, lasing occurs and, by most indications, hyperdimensional Torsion Energy is drawn from the vacuum potential of the space matrix.

BIBLIOGRAPHY

[**NOTE:** Back issues of *DNA Monthly*, where a number of articles listed below appear, can be read online at **http://www.potentiation.net**.]

Alexjander, Susan, "Music to the Ears: The Infrared Frequencies of DNA Bases" (*DNA Monthly*, September 2005)

Argüelles, José, *The Mayan Factor: Path beyond Technology* (Bear & Co., 1987)
— *Earth Ascending: An Illustrated Treatise on Law Governing Whole Systems* (Bear & Co., 1988)

Ascension Alchemy (Webpage on tetrahedral merkabah geometry) (http://www.asc-alchemy.com/chrys8.html)

Bailey, Alice, *Initiation Human and Solar* (Lucis Publishing Co., 1997)

Bartlett, Richard, *Matrix Energetics: The Science and Art of Transformation* (Atria Books, 2007)

Berendt, Joachim-Ernst, *The World Is Sound—Nada Brahma: Music and the Landscape of Consciousness* (Inner Traditions International, 1991)

Bischof, Marco, "Biophotons: The Light in Our Cells" (Webpage summarizing key points in the German book by the same name) (http://www.bibliotecapleyades.net/ciencia/ciencia_fuerzasuni verso06.htm)

Blavatsky, Helena, *Isis Unveiled* (Theosophical University Press, 1976)

Bohm, David, *Wholeness and the Implicate Order* (Routledge, 2002)

Booth, Robert, "Dust 'Comes Alive' in Space" (*UK Times Online* at http://www.timesonline.co.uk, August 2007)

Braden, Gregg, *Awakening to Zero Point: The Collective Initiation* (Radio Bookstore Press, 1997)
—*The Isaiah Effect: Decoding the Lost Science of Prayer and Prophecy* (Three Rivers Press, 2000)
—*The God Code: The Secret of Our Past, the Promise of Our Future* (Hay House, Inc., 2004)
—*The Divine Matrix: Bridging Time, Space, Miracles, and Belief* (Hay House, Inc., 2007)
—*The Spontaneous Healing of Belief: Shattering the Paradigm of False Limits* (Hay House, Inc., 2008)
—*Fractal Time: The Secret of 2012 and a New World Age* (Hay House, Inc., 2009)

Broe, Robert and Kerrie, *Absurd Medical Assumptions* (Tuberose Publishing, 1997)

Bryce, Sheradon, *Joy Riding the Universe: Snapshots of the Journey* (HomeWords Publishing, 1993)

Calleman, Carl Johan, *Solving the Greatest Mystery of Our Time: The Mayan Calendar* (Garev, 2001)
—*The Mayan Calendar and the Transformation of Consciousness* (Bear & Co., 2004)

Carey, Bjorn, "Cosmic 'DNA' Double Helix Spotted in Space," (http://www.space.com; excerpt reprinted in *DNA Monthly*, April 2006)

Carey, Ken, *The Starseed Transmissions* (HarperSanFrancisco, 1982)
—*Terra Christa* (Uni-Sun, 1985)
—*Return of the Bird Tribes* (Uni-Sun, 1988)
—*Starseed—The Third Millennium: Living in the Posthistoric World* (HarperSanFrancisco, 1991)

Champion, Joe, "Transdimensional Healing with the ADAM Technology" (*Nexus*, March-April 2004)

Chishima, Kikuo, *Revolution of Biology and Medicine* (Neo-Haematological Society Press, 1972)

Chopra, Deepak, *Ageless Body, Timeless Mind: The Quantum Alternative to Growing Old* (Harmony, 1995)

Clow, Barbara Hand, *The Pleiadian Agenda: A New Cosmology for the Age of Light* (Bear & Co., 1995)
—*Catastrophobia: The Truth Behind Earth Changes* (Bear & Co., 2001)

Cutler, Ellen, *Winning the War against Immune Disorders and Allergies: A Drug Free Cure for Allergies* (Delmar Thomson Learning, 1998)

Dossey, Larry, *Healing Words: The Power of Prayer and the Practice of Medicine* (HarperSanFrancisco, 1997)
—*Reinventing Medicine: Beyond Mind-body to a New Era of Healing* (HarperSanFrancisco, 1999)

Eisenstein, Charles, *The Ascent of Humanity: Civilization and the Human Sense of Self* (Panenthea Productions, 2007)

Elkins, Don, Rueckert, Carla and McCarty, James Allen, *The Law of One: Books I-V* (Whitford Press, 1984-98)

Emoto, Masaru, *The Hidden Messages in Water* (Beyond Words Publishing, 2004)

English, John, *The Shift: An Awakening* (Dreamtime Publications, 2004)

Fosar, Grazyna and Bludorf, Franz, *Vernetzte Intelligenz* ("Networked Intelligence") (Omega Verlag Bongart-Meier, 2001) (Currently unavailable in English. Visit the authors' website at http://www.fosar-bludorf.com.)

Free, Wynn with Wilcock, D., *The Reincarnation of Edgar Cayce?: Interdimensional Communication and Global Transformation* (Frog, Ltd., 2004)

Gariaev, Peter, "An Open Letter from Dr. Peter Gariaev, the Father of Wave-genetics" (*DNA Monthly*, September 2005)
—"A Brief Introduction to Wave-genetics: Scope and Possibilities" (*DNA Monthly*, April 2009)

Gardner, Laurence, *Genesis of the Grail Kings* (Fair Winds Press, 2002)

Garnett, Merrill, *First Pulse: A Personal Journey in Cancer Research* (First Pulse Projects, Inc., 1998)

Gerber, Richard, *Vibrational Medicine: New Choices for Healing Ourselves* (Bear & Co., 1988)

Gibbs, W. Wayt, "The Unseen Genome: Gems among the Junk" (*Scientific American*, November 2003)

Goldman, Jonathan, *Healing Sounds: The Power of Harmonics* (Healing Arts Press, 1992)

Goswami, Amit, *The Self-aware Universe: How Consciousness Creates the Material World* (Tarcher, 1995)

Grafio, Sai, Esoteric Articles on DNA (http://astromantic.arts.com)

Gray, William, *The Talking Tree* (Weiser, 1977)

Grof, Stanislov, *The Holotropic Mind: The Three Levels of Human Consciousness and How They Shape Our Lives* (HarperSanFrancisco, 1993)

Hawkins, David R., *Power vs. Force: The Hidden Determinants of Human Behavior* (Hay House, Inc., 1995)

Henry, William, *Godmaking: How Ancient Myths of DNA Reveal the Miracle Healing Power of Our Mystic Anatomy* (Scala Dei, 2000)
—*The Healing Sun Code: Rediscovering the Secret Science and Religion of the Galactic Core and the Rebirth of Earth in 2012* (Scala Dei, 2001)
—*The Language of the Birds: Our Angelic Connection* (Scala Dei, date unspecified)

Horowitz, Leonard G., *Emerging Viruses: AIDS and Ebola— Nature, Accident or Intentional?* (Tetrahedron, 1996)
—*DNA Pirates of the Sacred Spiral* (Tetrahedron, LLC, 2004)

Horowitz, Leonard G. and Puleo, Joseph S., *Healing Codes for the Biological Apocalypse* (Tetrahedron, 1999)

Huggins, Hal, *It's All in Your Head: The Link between Mercury Amalgams and Illness* (Avery Publishing Group, 1993)

Hunt, Valerie, *Infinite Mind: Science of the Human Vibrations of Consciousness* (Malibu Publishing Co., 1989)

Hurtak, J. J., *The Book of Knowledge: The Keys of Enoch* (Academy for Future Science, 1977)

Jahn, Robert G. and Dunne, Brenda, *Margins of Reality: The Role of Consciousness in the Physical World* (Harcourt Brace Jovanovich, 1987)

Jenkins, John Major, *Maya Cosmogenesis 2012* (Bear & Co., 1998)
—*Galactic Alignment: The Transformation of Consciousness according to Mayan, Egyptian, and Vedic Traditions* (Bear & Co., 2002)

Jernigan, David A. "Illuminated Physiology and the Medical Uses of Light" (*DNA Monthly*, May 2009)

Kelleher, Colm A., "Retrotransposons as Engines of Human Bodily Transformation" (*Journal of Scientific Exploration*, Spring 1999)

LaViolette, Paul, *Earth Under Fire: Humanity's Survival of the Ice Age* (Bear & Co., 2005)

Linsteadt, Stephen, "Frequency Fields at the Cellular Level" (*DNA Monthly*, November 2005)

Lipton, Bruce, *An Introduction to the Biology of Consciousness* (videotape) (CELL, 1995)
—*The Biology of Belief: Unleashing the Power of Consciousness, Matter, and Miracles* (Mountain of Love/Elite Books, 2005)

Marciniak, Barbara, *Bringers of the Dawn: Teachings from the Pleiadians* (Bear & Co., 1992)
—*Earth: Pleiadian Keys to the Living Library* (Bear & Co., 1995)
—*Path of Empowerment: Pleiadian Wisdom for a World in Chaos* (Inner Ocean Publishing, 2004)

McKenna, Terrence, *True Hallucinations* (HarperCollins, 1994)

McTaggart, Lynne, *The Field: The Quest for the Secret Force in the Universe* (Quill, 2003)

Miller, Neil Z., *Immunization: Theory vs. Reality* (New Atlantean Press, 1996)

Miller, Iona and Miller, Richard A., "From Helix to Hologram: An Ode on the Human Genome" (*Nexus*, September-October 2003; reprinted in *DNA Monthly*, October 2005)
—"The Universe Is Obsolete: A Gallery of Multiverse Theories" (*DNA Monthly*, July-August 2005)

Miller, Richard A., Miller, Iona and Webb, Burt, "Quantum Bioholography: A Review of the Field from 1973-2002" (*The Journal of Non-Locality and Remote Mental Interactions*, October 2002)

Miller, Richard A., Webb, Burt and Dickson, Darden, "A Holographic Concept of Reality" (*Psychoenergetic Systems*, Vol. 1, 1975; reprinted in *DNA Monthly*, May 2007)

Mohr, Bärbel, "DNA's Hypercommunication: The 'Living Internet' inside of Us" (2002) (Summary of the German book *Vernetzte Intelligenz* ["Networked Intelligence"] by Grazyna Fosar and Franz Bludorf) (http://www.bibliotecapleyades.net/ciencia/ciencia_genetica02 .htm)

Motoyama, Hiroshi, *Theories of Chakras: Bridge to Higher Consciousness* (Quest, 1981)

Nambudripad, Devi, *Say Goodbye to Illness* (Delta Publishing Co., 1999)

Narby, Jeremy, *The Cosmic Serpent: DNA and the Origins of Knowledge* (Jeremy P. Tarcher/Putnam, 1998)

Opitz, Christian, "Enlightenment and the Brain" (*DNA Monthly*, May 2006)

"Partial Ingredients for DNA and Protein Found around Star" (http://legacy.spitzer.caltech.edu/Media/releases/ssc2005-26)

Pearce, Joseph C., *Evolution's End: Claiming the Potential of Our Intelligence* (HarperCollins, 1992)
—*The Biology of Transcendence: A Blueprint of the Human Spirit* (Inner Traditions, 2001)

Polich, Judith B., *Return of the Children of Light: Incan and Mayan Prophecies for a New World* (Bear & Co., 2001)

Pinchbeck, Daniel, *2012: The Return of Quetzalcoatl* (Jeremy P. Tarcher/Penguin, 2006)

Pribram, Karl, *Languages of the Brain* (Prentice-Hall, Inc., 1971)

Radhoff, Ron (quoted in *New Science News*, Vol. III, No. 2, p. 7)

Redfield, James, Murphy, Michael and Timbers, Sylvia, *God and the Evolving Universe: The Next Step in Personal Evolution* (Jeremy P. Tarcher/Putnam, 2002)

Rein, Glen, "Effect of Conscious Intention on Human DNA" (Proc.Internat.Forum on New Science, 1996)

Ruiz, Miguel, *Beyond Fear: A Toltec Guide to Freedom and Joy* (Council Oak Books, 1997)
—*The Four Agreements: A Practical Guide to Personal Freedom, A Toltec Wisdom Book* (Amber-Allen Publishing, 2001)

Sahtouris, Elisabet, "Living Systems in Evolution" (*DNA Monthly*, April 2007)

Sheldrake, Rupert, *The Presence of the Past: Morphic Resonance and the Habits of Nature* (Inner Traditions, 1995)

Sitchin, Zecharia, *The Cosmic Code* (Avon Books, 1998)

Smelyakov, Sergey, "The Auric Time Scale and the Mayan Factor: Demography, Seismicity and History of Great Revelations in the Light of the Solar-planetary Synchronism" (Kharkov, 1999)

Smolin, Lee, *The Life of the Cosmos* (Oxford University Press, 1999)

"Surfactants" (Webpage on the use of light in cellular communication and zero point energy from lasing) (http://www.miracleii.com/surfactants-ezp-13.html)

Tachí-Ren, Tashíra, *What Is Lightbody?* (New Leaf Distributing, 1990)

Talbot, Michael, *The Holographic Universe* (HarperPerennial, 1992)

Tansley, David V., *Radionics and the Subtle Anatomy of Man* (The C. W. Daniel Company Ltd., 1972)
—*Radionics: Interface with the Ether Fields* (The C. W. Daniel Company Ltd., 1975)

Tolle, Eckhart, *The Power of Now: A Guide to Spiritual Enlightenment* (New World Library, 1999)

Vonderplanitz, Aajonus, *We Want To Live* (Carnelian Bay Castle Press, LLC, 1997)

Ward, Geoff, *Spirals: The Pattern of Existence* (Green Magic, 2006)

Weinhold, Barry K. and Janae B., "Preparing for the Shift" (multimedia presentation, 2004)
—"Finding the Holy Grail" (*Inner Tapestry*, December-January 2004)

White, John and Krippner, Stanley, *Future Science: Life Energies and the Physics of Paranormal Phenomena* (Anchor Books, 1977)

Wilcock, David, *The Divine Cosmos* (http://www.divinecosmos.com)
—"The Ultimate Secret of the Mayan Calendar: An Imploding Cycle of Energy Increase, Culminating in 2012-2013 A.D." (http://www.scottmandelker.com; reprinted in *DNA Monthly*, January 2006)
—"Kozyrev: Aether, Time and Torsion" (http://www.divinecosmos.com; reprinted in *DNA Monthly*, May 2008)
—Personal website and weblog (http://www.divinecosmos.com)

INDEX

oneness (Oneness), 188, 215, 233, 249
operating system, 61, 155, 206, 207
Ophiuchus, 133, 270, 291
opportunity, 111, 145, 168, 206
organism, 40, 60, 72, 73, 77, 80, 87, 88, 94, 97, 128, 129, 158, 159, 173, 205, 221, 222, 251, 268, 279, 298
original (Original)
sin, 88
Wound, 88, 93, 100, 209
oxygen, 55, 79, 132, 152
oxygenation, 252

P

pain, 38, 222, 230, 236, 238, 243, 247
palindrome, 70
Pangaea, 117, 284
paradigm, 34, 57, 59, 62, 95
paradox, 193, 195, 226
paranoia, 124, 274
parasite (Parasite), 38, 92, 222, 284
parietal lobes, 187, 188
Parkinson disease, 44, 270
particle-wave, 88, 204
Path
of Nature, 174, 251
of Technology, 174, 251
pathogen, 40, 41, 45
peace, 168, 227, 238, 274, 277
perception, 115, 151, 175, 215, 290
perfection, xvii, 209
periodontal disease, 274
permanent atom, 223, 269, 270, 271, 272, 273, 274
Atmic, 223, 278
pharmaceutical, 63, 65, 237, 274
Phi, 81, 125, 126, 134, 277, 285, 294
philosophy, xx, 198, 204, 208, 209
phoenix, 182
phonon, 75, 77, 285

photography
Kirlian, 52, 151
radionic, 108
photon (Photon), 75, 77, 80, 144, 196, 203, 207, 285
Band (Belt), xi, 122, 123, 124, 125, 126, 128, 130, 131, 139, 149, 162, 225, 279, 281, 284, 290, 291, 294, 295, 297
photosynthesis, 162
phototherapy, 163
Pistis Sophia, 184
placebo effect, 34
plasma, 81, 212, 278, 291
poem (poetry), 60, 87, 251
polarization, 115
Popol Vuh, 67
positron, 207, 225, 291
positronium, 206, 207, 291
possession, 274
Potentiation (Electromagnetic Repatterning), vi, xviii, xix, xx, 31, 35, 64, 98, 99, 100, 101, 102, 105, 106, 107, 108, 110, 111, 168, 169, 171, 177, 208, 223, 229, 230, 231, 232, 233, 235, 236, 237, 238, 239, 240, 242, 245, 252, 263, 264, 267, 268, 269, 278, 282, 283, 284, 288, 292, 293, 294
potentiator, 100, 268, 269
power, xviii, xx, 32, 34, 54, 59, 62, 87, 141, 168, 172, 173, 174, 175, 191, 204, 205, 228, 234, 251
prana, xxiv, 96, 293
prayer, xx, 31, 32, 33, 63, 174, 226, 255, 257, 258, 262, 277
active, 32, 226, 277
precession (of the equinoxes), 116
prefrontal lobes, 102, 187, 189, 282
Princeton Engineering Anomalies Research (PEAR), 35
prion, 41
propaganda, 173, 254

protein, 66, 68, 70, 73, 80, 172,
293, 304
synthesis, 64, 65, 75, 88,
293
psyche, 140
psychiatrist, 139
psychiatry, 34
psychic surgery, 38
psychoanalysis, 34
psychologist, 34, 115
psychology, 151, 179
punctuated equilibrium, 121

Q

qigong, 43
quantum
biocomputer, 86, 158
bioholography, 77, 96, 304
biology, 51, 156, 227
connection, 140
defined, 292
hologram, 77
holography, 140
medium, 94
outcome, 32, 226, 227, 277
physics, 33, 95
potential, 34, 281, 292
quest, 213
Quetzalcoatl, 185, 186, 219,
297, 304

R

Ra, 67, 220, 224
radiance, 94, 185
radiation, xxiii, 35, 63, 65, 80,
85, 88, 125, 172, 203, 292
radio
telescope, 212
wave, 57, 85, 86
radionics, 38, 57, 59, 89, 108,
109, 259, 260, 292, 306
rainbow, 181, 193, 292
rash, 111, 237, 240, 272
ray (true color), 193, 219, 269,
270, 271, 272, 273, 274, 292
reality
adjusting, 228
biological, 155, 161, 198
experience of, 77, 102

parallel, 122, 145, 152, 227
precipitated, 77
rebuilding, 102, 236
receding gums, 38, 274
reflexology, 241
regeneration, 73, 163
Regenetics (Method)
as a synonym for
electrogenetics and
wave-genetics, 293
defined, 293
Forum, 252
journal, 231, 257, 262
model, 143, 203, 297
session, 63, 251, 254, 256,
257
rehydration, 163, 294
reiki, 38, 57, 241, 260
religion, vi, 66, 88, 115, 168,
210, 212
remote viewing, 95
replication, 71, 77, 176, 283
reprogramming, 69, 296
respiration, 87
Resurrection, 145, 180
retrotransposon, 130, 157, 303
retrovirus, 39, 40, 41
reunion, 206
reverse
transcriptase, 41
transcription, 40, 65, 66,
68, 70, 71, 72, 73, 172
revolution, i, xvii, 78, 85, 86,
111, 137, 201, 204, 220, 256,
298, 300
rewriting, 69, 129, 296
ribosome (Ribosome), 66, 293
RNA (mRNA), 40, 66, 68, 70,
71, 72, 73, 80, 88, 130, 172,
183, 221, 227, 271, 277, 283,
291, 292, 293, 296
Rolfing®, 108
Roman (Catholic) Church, 92,
160, 171, 179, 180, 251

S

sacred (Sacred)
geometry, 60, 90, 149, 285
marriage, 100, 217

ABOUT THE AUTHOR

Sol Luckman is an acclaimed author of fiction and nonfiction and pioneering ink painter whose work has appeared on mainstream book covers. His books include the international bestselling *Conscious Healing: Book One on the Regenetics Method* and its popular sequel, *Potentiate Your DNA: A Practical Guide to Healing & Transformation with the Regenetics Method*. His visionary novel, *Snooze: A Story of Awakening*, won the 2015 National Indie Excellence Award for New Age Fiction. *Snooze* further proved its literary merit by being selected as a 2016 Readers' Favorite International Book Award Finalist in the Young Adult-Coming of Age category and receiving an Honorable Mention in the 2014 Beach Book Festival Prize Competition in the General Fiction category. Sol's latest book, *The Angel's Dictionary: A Spirited Glossary for the Little Devil in You*, winner of the 2017 National Indie Excellence Award for Humor, reinvigorates satire to prove that—though we might not be able to change the world—we can at least have a good laugh at it. Then again, maybe laughter can transform the world! View Sol's paintings, read his blog and learn more about his work at **www.CrowRising.com**.

The first DNA activation in the "revolutionary healing science" (*Nexus*) of the Regenetics Method, Potentiation employs special linguistic codes—produced vocally and mentally—to stimulate a self-healing and transformational ability in DNA.

In this masterful exploration of sound healing by bestselling author Sol Luckman, learn how to activate your genetic potential—in a single, thirty-minute session!

Besides teaching you a leading-edge technique you can perform for your family, friends and even pets, *Potentiate Your DNA* also:

1. Provides a wealth of tried and true supplemental tools for maximizing your results; and
2. Outlines a pioneering theory linking genetics, energy and consciousness that is sure to inspire alternative and traditional healers alike.

Potentiate Your DNA "is both fascinating and an astounding, perhaps even world-changing theory." —*New Dawn* Magazine

"*Potentiate Your DNA* is brilliant and cutting-edge. Luckman has succinctly and elegantly provided a comprehensible intellectual framework for understanding the profound role of DNA in healing and transformation." —Brendan D. Murphy, author of *The Grand Illusion*

"If you love the cutting-edge of the cutting-edge ... read this book!" —Dr. David Kamnitzer

"The work defined in this book and Sol Luckman's previous book, *Conscious Healing*, should be the starting place of every health practice." —Dr. Julie TwoMoon

Learn more at **www.PhoenixRegenetics.org**.

CPSIA information can be obtained
at www.ICGtesting.com
Printed in the USA
BVHW080431020921
615808BV00002B/153